THE PLATONIC SOLIDS ACTIVITY BOOK

VISUAL GEOMETRY PROJECT

Ann E. Fetter
Cynthia Schmalzried
Nancy Eckert

Doris Schattschneider
Eugene Klotz

Funding for the Visual Geometry Project has been provided by a grant from the National Science Foundation.

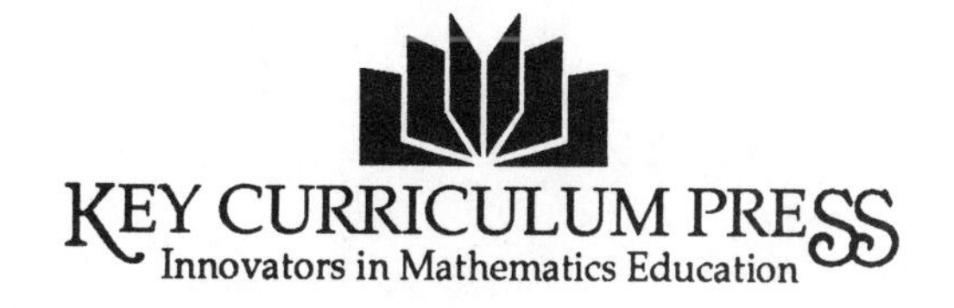

Written and edited by the Visual Geometry Project: Nancy Eckert, Ann E. Fetter, Dr. Doris Schattschneider, Cindy Schmalzried

Assisted by: Barry Peratt and Sara McGill Mace

Further editing by Key Curriculum Press: Steve Rasmussen, Dan Bennett

Layout and Graphics by Ann E. Fetter

Cover by John Odam Design

Many thanks to the teachers and students who used these materials in their preliminary form. Your classroom experience and comments were invaluable in keeping this project grounded in reality.

Limited Reproduction Permission

Published by Key Curriculum Press, P.O. Box 2304, Berkeley, CA 94702.

National Science Foundation Grant No. MDR-8850726

Printed in the United States of America

ISBN 1-55953-043-X

TABLE OF CONTENTS

TEACHER'S GUIDE

BLACKLINE MASTERS

TEACHER'S GUIDE: Introduction, Terminology, Glossary, Annotated Activities with Teacher Comments, and References

Introduction

The Platonic solids have fascinated mathematicians and engineers, artists and architects, philosophers and astronomers for thousands of years. Why are they so interesting? *The Platonic Solids* materials have been designed to help students formulate some answers to that question. These materials emphasize the beauty and perfection of these deceptively simple shapes, shapes which so impressed Plato that he used them to describe his vision of a perfect world.

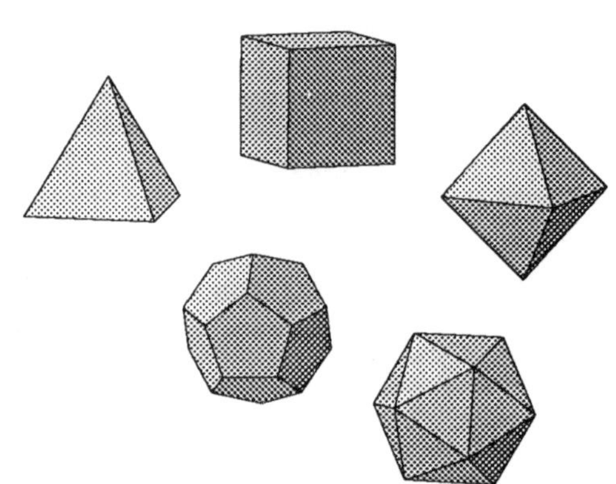

What are the Platonic solids? They are also known as regular polyhedra because they are regular, or uniform, in every aspect. All the faces of a regular polyhedron are congruent regular polygons, and the same number of faces meet at each vertex in exactly the same way. It may seem that such regularity would be easy to achieve, but in fact it is rare. There are only five regular polyhedra, only five solids which stand out from all other three-dimensional figures in their perfection, their symmetry, and their elegance.

Analysis of the regular polyhedra reveals a myriad of connections to two-dimensional figures, to other three-dimensional shapes, and to each other. Students quickly discover many beautiful and intriguing patterns. But the Platonic solids are not of interest solely to mathematicians. These figures appear in art and in nature, in the shape of crystals, molecules, and living organisms, in simple drawings and intricate sculptures.

The Platonic Solids Materials

The Platonic Solids materials were developed as part of the Visual Geometry Project, a National Science Foundation-funded project directed by Dr. Eugene Klotz at Swarthmore College and Dr. Doris Schattschneider at Moravian College in Pennsylvania. The goal of the Visual Geometry Project was to develop supplementary materials that would enable high school geometry students to improve spatial visualization skills. Traditional textbook instruction does little to address these skills; most geometry texts are especially lacking when it comes to teaching three-dimensional visual reasoning skills. The Visual Geometry Project developed two units, *The Platonic Solids* and *The Stella Octangula*, that employ video and hands-on manipulatives to give students experience with visual reasoning in ways textbooks cannot. *The Geometer's Sketchpad*, software for doing dynamic geometry on Macintosh computers, was also developed by the Visual Geometry Project.

The complete set of Platonic solids materials consists of a computer-animated videotape, an activity book with blackline masters, and a manipulative kit to make model-making easy. Each of these three components is described in more detail below. The materials were designed for use in a high school geometry class, but many activities are appropriate for younger students, and the materials have also been used in pre-service and in-service courses for teachers. You'll have to choose what activities to do in your class depending on how much time you want to spend. The complete unit, including time to show the videos and do some independent student projects, will take about three weeks to complete. You can do the activities all at once, or do a few at a time and return to them throughout the year.

The Platonic Solids Videos

The Platonic Solids is an 18-minute video which uses the power of computer animation to show students properties of regular polyhedra which are difficult to visualize from blackboard pictures or even static models. Beginning with a demonstration that exactly five regular polyhedra exist and a discussion of the history and uses of solids, the video goes on to portray the beautiful projective views of each wire-model shape, followed by a presentation of the dual relationship of pairs of Platonic solids.

The video is designed for flexibility in the classroom. You can show it all at once or in parts. It may be useful for students to do some activities with the shapes before viewing the video. Students discover the fact that there are exactly five regular polyhedra in the first activity, so don't show the video before your students do that activity unless you want to give away the answers. You'll probably want to show the video more than once. In the teacher comments we have provided suggestions for using the video in conjunction with the activities. Those suggestions are summarized below, but you are encouraged to follow your own ideas, too.

First showing:

After Activity 1: The Regular Polyhedra.

Show the entire video, or just the first part that shows different views of the Platonic Solids and demonstrates why there are exactly five.

Second showing:

After Activity 6: Duality of Polyhedra.

Show the part that demonstrates duality.

Third showing:

At the end of the unit.

Show the entire video to summarize and bring closure to the unit.

A video for teachers, *Using the Platonic Solids Materials*, follows *The Platonic Solids* on the same tape. This 15-minute video suggests some ways in which the materials can be used in the classroom, gives some useful factual background, and shows scenes of actual classroom use.

The Platonic Solids Activity Book

The Platonic Solids Activity Book provides a hands-on, visual approach to the study of geometry. Students build three-dimensional models and carefully analyze their properties. The exercises encourage exploration and spontaneous discovery of important geometry facts. The activities also help students see how relationships in two-dimensional geometry can be extended to three dimensions.

The activity book consists of three preliminary activities concerning polygons and eight activities dealing with the Platonic solids. The student activities pages are blackline masters that can be reproduced for students. Each activity is designed to take approximately one class period. Specific instructions for building the models are included with each activity. Students may work in pairs or small groups. The annotated activities and teacher comments give an overview of each activity, answers to the exercises, additional factual information, and suggestions for further discussion. A page of terminology that students are expected to know prior to doing the activities is included in the blackline masters and in the Teacher's Guide, with comments. The Teacher's Guide and Blackline

Masters also include a graphical glossary of new terms encountered in the activities. You may want to reproduce the terminology and glossary pages for students as well as the activities.

Also included in the activity book are 14 student projects, which are intended for independent work by students or groups of students. These student projects vary greatly in difficulty; the background knowledge required for each one is clearly indicated. References are suggested where appropriate.

References listed in the student projects are collected in a more complete list of references found in the Teacher's Guide. The purpose of this references list is to lead teachers and students working on projects to resources for further study of the Platonic solids. These books need not be available to complete the projects, but they provide opportunities to explore topics in greater depth. You might want to pull the references out of the activity book, or make an extra copy or two to make available to students. Most of these references are not likely to be available in a school library—you'll probably need to look for them in a university library. You can make the trip to the library yourself to check out a collection to keep in your classroom while your class is studying the Platonic solids, or send individual students to do research on their own. Over time you may want to convince your school library to obtain some of these books, or you may choose to obtain some for your professional library.

Building the Models

Materials for the models required in the activities are available from two sources: *The Platonic Solids Activity Book* and *The Platonic Solids Manipulative Kit* (see below for a complete description of the kit). The activity book contains blackline masters of student activity sheets and panel pages and net pages for all the models in the activities and student projects. Everything you need is in the activity book, if you want to invest the extra time required to create panels from the blackline masters. Be warned though, that teachers could spend many hours preparing the panels and nets in advance, even just enough for eight groups of students in one class. For students to make them during class time would require two or more class periods. The manipulative kit can save a considerable amount of time over creating the models from scratch with the blackline masters.

If you're using the blackline masters in the activity book, make models by transferring the panels from the masters onto heavier paper (oak tag or manila file folders work well). The panels can be transferred by direct photocopying or by making a ditto master. Seven copies of each of the two panels net pages will give you enough polygon panels for one group to complete all the activities. To make models without edge lines, hold a copy of the blackline panel on top of your paper. Using a pin or compass point, poke a tiny hole through each vertex on the panel. This marks the vertices on the heavy paper, and then the edges can be scored and cut. To score the edges and the fold lines, use the point of a compass held against a straightedge. For models that must be glued, such as the tetrahedron and cube corner nets for Activity 7, use rubber cement, liquid glue sticks, or tape. For polygon panels, which will be attached with rubber bands, use a hole puncher to make an indentation at each corner. Some activities call for a transparent cube. Transfer the transparent cube net onto a piece of 10 mil. acetate using the same method described above for oak tag. (Acetate is available in art and hobby supply stores and catalogs and in some office supply stores and catalogs.) Tape the cube edges with transparent tape. Edges may break when scored and folded, but can be held together with tape.

The materials and models required for each activity are listed in the teacher comments.
That information is summarized in the chart on the following page.

Activity	Models to Construct	Panels/Materials for One Group of Students	Completed Models
P-2	None	1 Triangle, 1 Square, 1 Pentagon, 1 Rhombus	None
P-3	Tessellations	12 Triangles, 9 Squares, 9 Rhombuses, 2 Hexagons	None
1	Five Platonic Solids	32 Triangles, 6 Squares, 12 Pentagons	None
2	Transparent Cube	Acetate Cube Net, Straws, Rubber Bands	Five Platonic Solids
3	Pyramid, Prism, Cuboctahedron, Rhombic Dodecahedron	14 Triangles, 15 Squares, 2 Pentagons, 12 Rhombuses, Rubber Bands	None
4	None	None	All Models from Activities 2 and 3
5	None	None	Same as Activity 4
6	None	None	Octahedron, Transparent Cube
7	Optional: Tetrahedron, Four Cube Corners	Optional: Tetrahedron Net, Cube Corner Nets, Non-Permanent Marker, Rice, Birdseed, or Similar Material	Transparent Cube
8	None	**Rubber Bands, Non-Permanent Marker, Optional: Rice, Birdseed, or Similar Material**	Transparent Cube

The Platonic Solids Manipulative Kit

You may also purchase a classroom model-making kit consisting of polygon panels, rubber bands, straws, and die-cut nets for assembling transparent acetate cubes. This classroom kit contains enough materials for eight groups of students to do all the activities, excluding the student projects, with a few panels left over. Panels used in Activities P-1 through P-3 should be kept and used to make models in later activities. The nets for the student projects and optional activities are available as blackline masters only; students must transfer these onto heavier paper. Polygon panels for the student projects can be obtained by taking apart models made in the activities and reusing their panels, or by creating new panels with the blackline masters. At the end of the unit, or the school year, you may want to have students take apart the models so you can reuse the panels with future classes.

Putting it All Together

The activity book and manipulatives are an important component of the Platonic solids materials. Students will enjoy the video, but it cannot actively engage them in the doing of mathematics. The manipulation involved in constructing the three-dimensional models facilitates learning in a way that cannot be achieved by listening or watching. By engaging students with the video, the activities, and the manipulatives, you'll address the greatest variety of individual learning styles of your students. We encourage you to make time in your schedules for students to build and analyze their own creations; we believe it will be time well-spent.

Terminology

Here are the most common terms used to describe two- and three- dimensional figures.

Polygons

Polygons are composed of connected line segments, called **edges**, which enclose a single region of the plane. Edges are also called the "sides" of the polygon. A **vertex** of a polygon is a point where exactly two edges meet, forming an **interior angle**.

The names of polygons describe how many edges they have:

triangle: three edges
quadrilateral: four edges
pentagon: five edges
hexagon: six edges
septagon: seven edges
octagon: eight edges
***n*–gon**: *n* edges

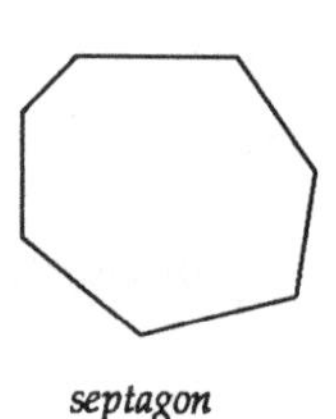

septagon

A polygon is **regular** if all of its edges have equal length and all of its interior angles have equal measure. A regular triangle is called an **equilateral triangle**, and a regular quadrilateral is called a **square**.

regular

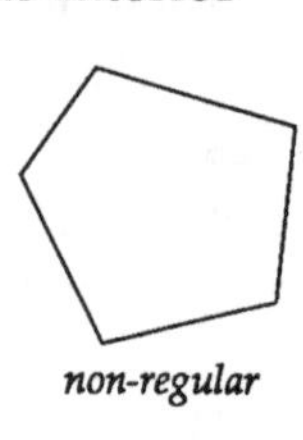

non-regular

A polygon is **convex** if any two points on its edges can be connected by a line segment which lies entirely inside the polygon. Otherwise, the polygon is **non–convex**.

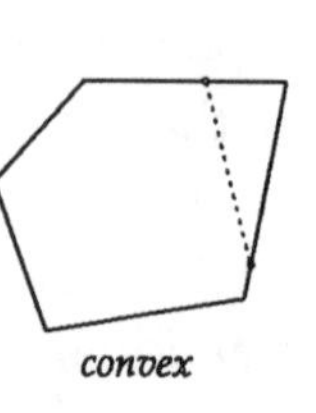

convex

non-convex

Polyhedra

Polyhedra are composed of polygons, called **faces**, which enclose a single region of space. An **edge** of the polyhedron is formed where exactly two faces are joined. A **vertex** of a polyhedron is a point where three or more edges meet.

The names of polyhedra describe how many faces they have:

tetrahedron: four faces
hexahedron: six faces
octahedron: eight faces
***n*–hedron**: *n* faces

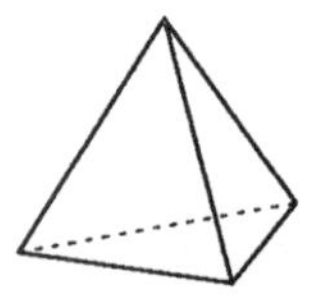

regular tetrahedron

A **pyramid** is formed when all vertices of a polygon are joined by line segments to a single point not in the same plane.

pentagonal pyramid

A polyhedron is **regular** if all of its faces are congruent regular polygons, and the same number of faces meet at each vertex in exactly the same way. A regular hexahedron is called a **cube**.

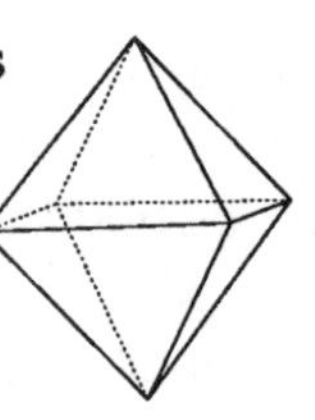

regular octahedron

A polyhedron is **convex** if any two points on its surface can be connected by a line segment which lies entirely inside or on the polyhedron. Otherwise it is **non–convex**.

convex

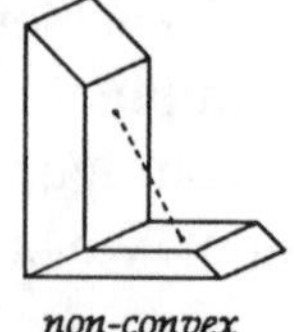

non-convex

Teacher Comments Terminology

Texts and references on polygons and polyhedra do not all agree on the definitions of these geometric forms. The terminology, descriptions, and definitions given here are commonly accepted and are precise enough to allow accurate descriptions of properties and relationships. All polyhedra that are commonly studied satisfy these definitions, and the restrictions stated insure that the figures possess certain expected important properties.

It is a worthwhile class exercise to have students attempt to formulate their own definitions of polygons and polyhedra. (Begin by using a dictionary to trace the meaning of the names.) In doing this exercise, students become aware of what they assume even though these assumptions are not explicitly stated, and why careful definitions are important. Ask for, and then examine several examples of two- and three-dimensional figures that might qualify as polygons or polyhedra based on their definitions. Ask if they would want to allow or exclude them and why. Here are a few two-dimensional figures that might be judged:

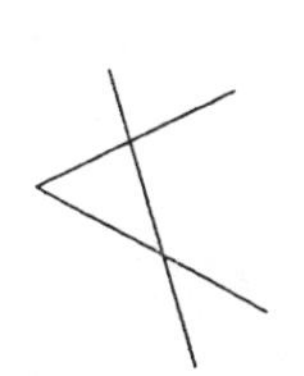

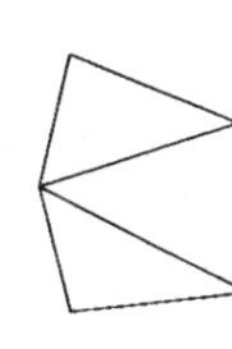
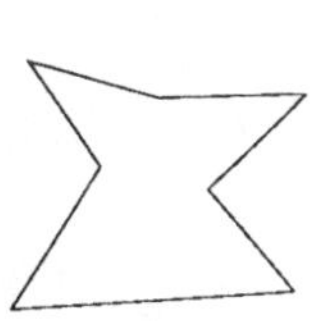
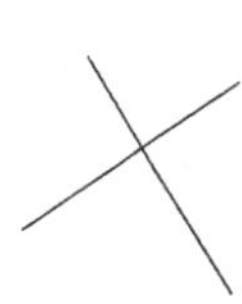
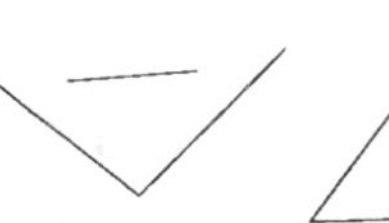
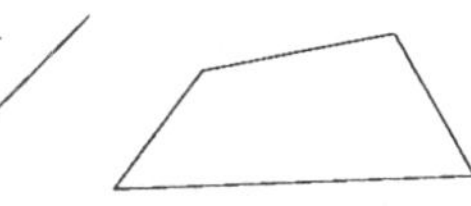

Comments on Descriptions

A standard convention is that an interior angle of a polygon is not a straight angle. A vertex of a polyhedron is also a vertex of each of the polygon faces which meet there. Thus it could also be defined as a point where three or more faces of the polyhedron meet. Another important property of polygons and polyhedra is that each edge joins exactly two vertices.

The two-column format of the page of terminology draws attention to the analogies between two-dimensional polygons and three-dimensional polyhedra. Here are some other analogies that can be pointed out:

A **triangle** is formed when both endpoints of a line segment (called the base) are joined by edges to a single point not collinear with the base.

A **triangle** is a *minimal polygon*, that is, it has the minimum number of edges (3) and minimum number of vertices (3) possible for a polygon.

Every **triangle** is convex.

If the edges of a **triangle** are of equal length, then the triangle is regular.

A **pyramid** is formed when all vertices of a polygon (called the base) are joined by edges to a single point not in the plane of the base.

A **tetrahedron** is a *minimal polyhedron*, that is, it has the minimum number of faces (4), minimum number of edges (6), and minimum number of vertices (4) possible for a polyhedron.

Every **pyramid** is convex.

If the edges of a **tetrahedron** are of equal length,

Note: The last two statements in the parallel columns above are not true if the word "triangle" is replaced by "polygon," or the words "pyramid" or "tetrahedron" are replaced by the word "polyhedron."

Glossary

The Platonic Solids

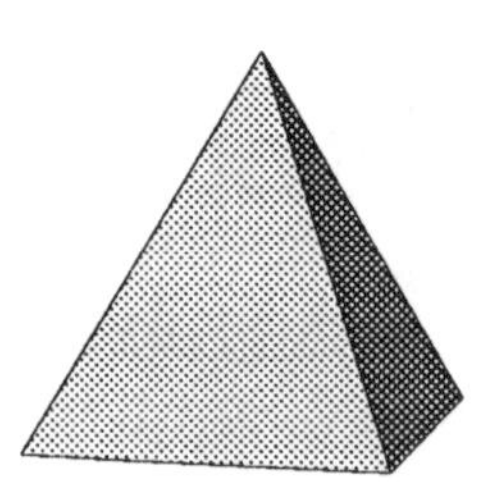

Tetrahedron

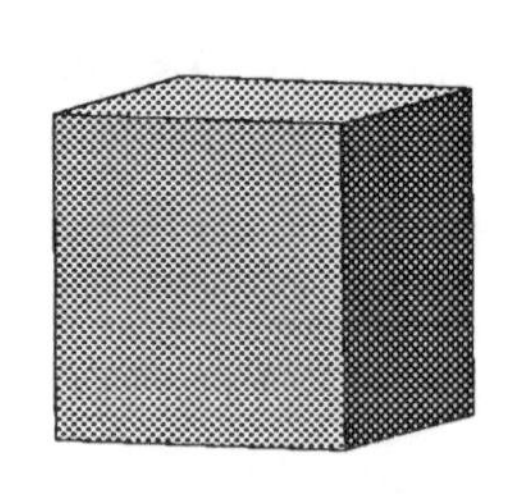

Cube

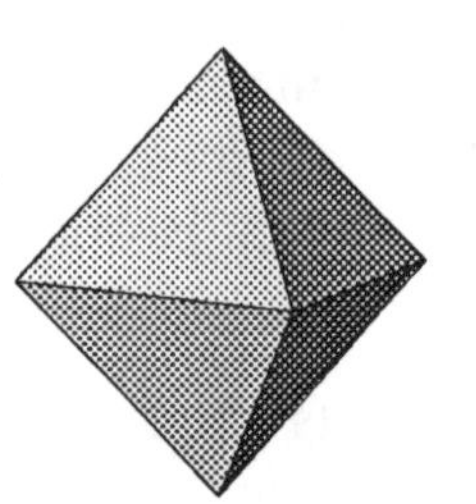

Octahedron

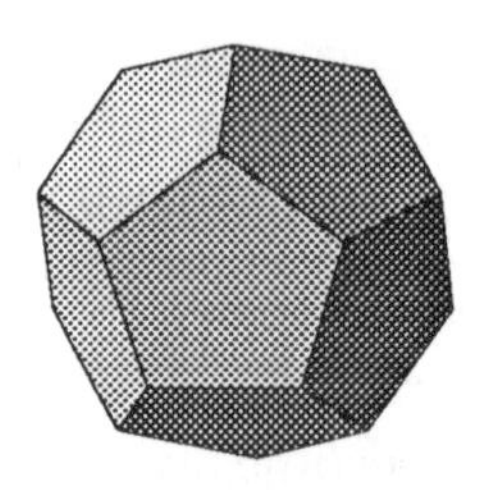

Dodecahedron

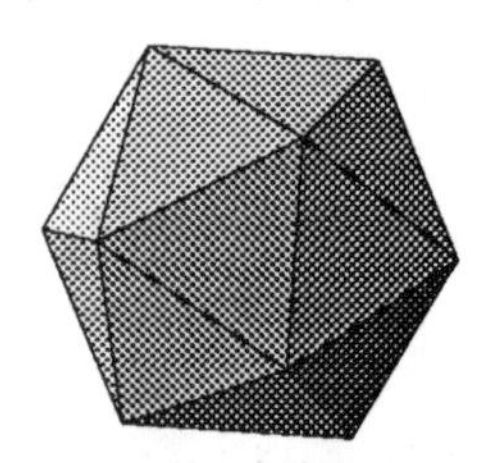

Icosahedron

The Archimedean Solids

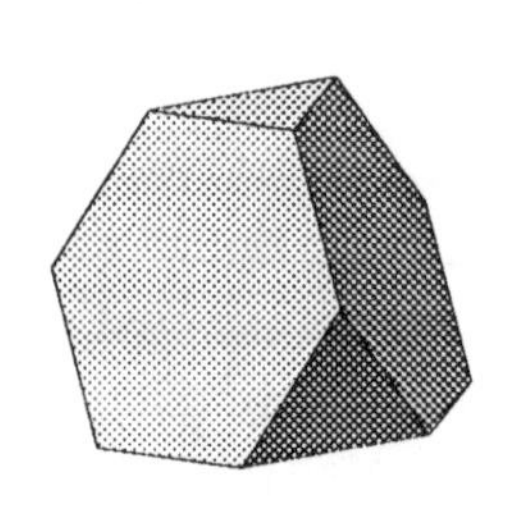

Truncated tetrahedron

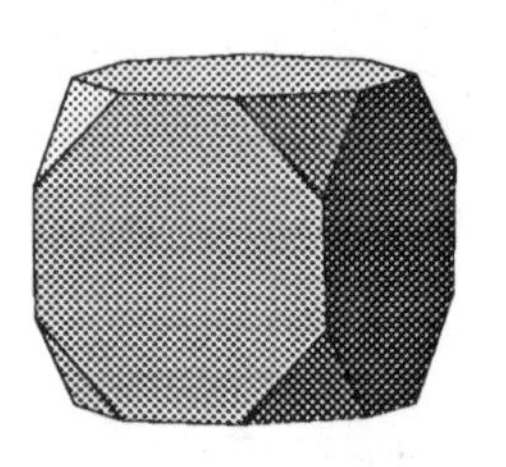

Truncated cube

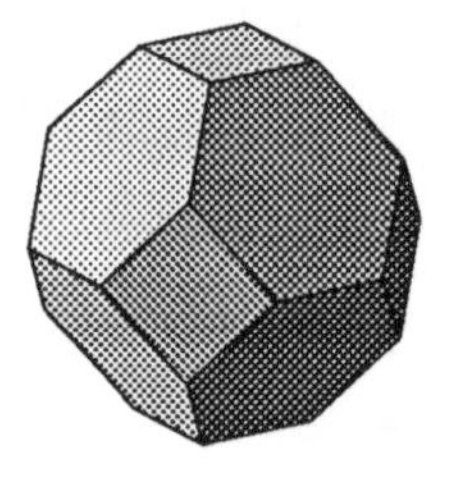

Truncated octahedron

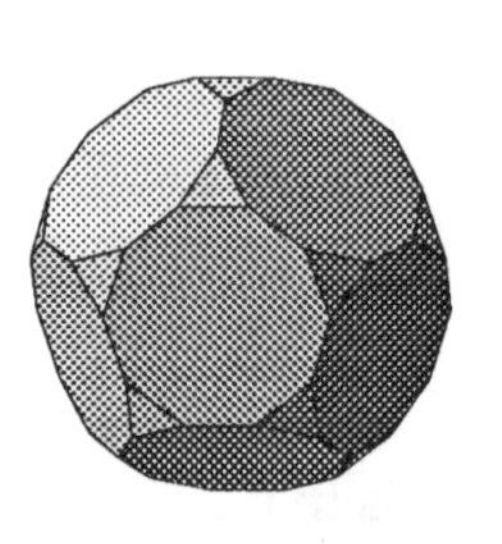

Truncated dodecahedron

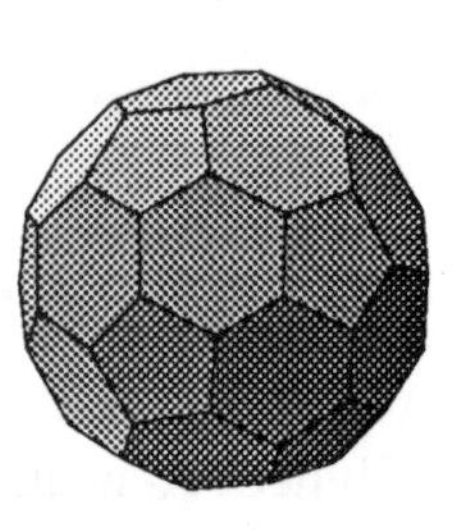

Truncated icosahedron

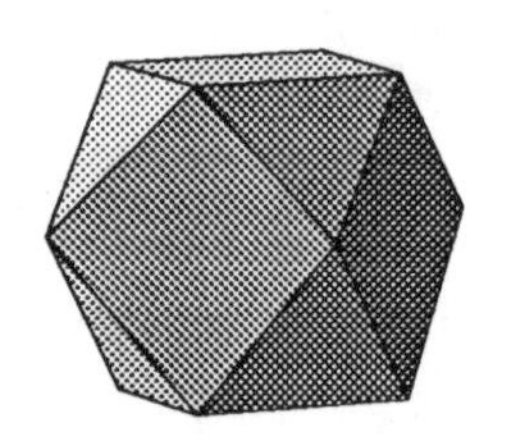

Cuboctahedron

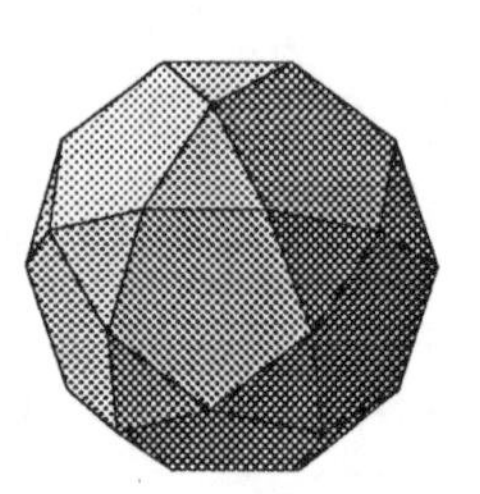

Icosidodecahedron

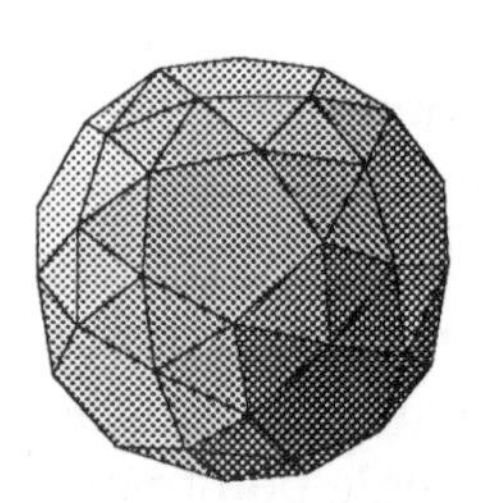

Snub dodecahedron

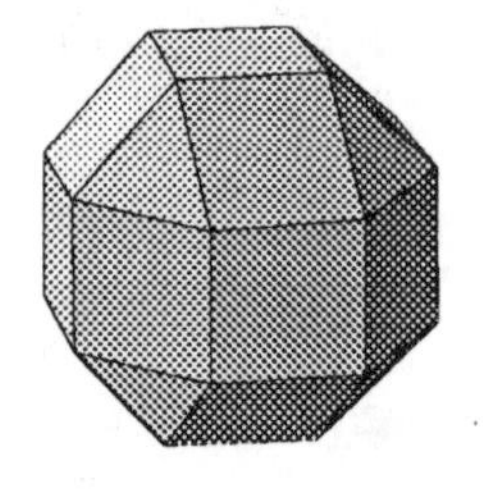

Rhombicuboctahedron

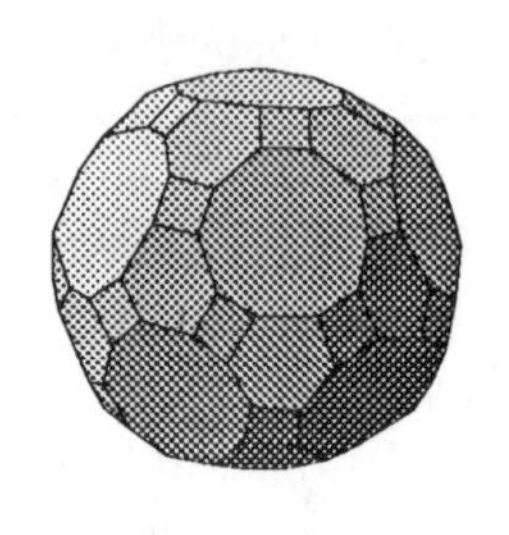

Great rhombicosidodecahedron

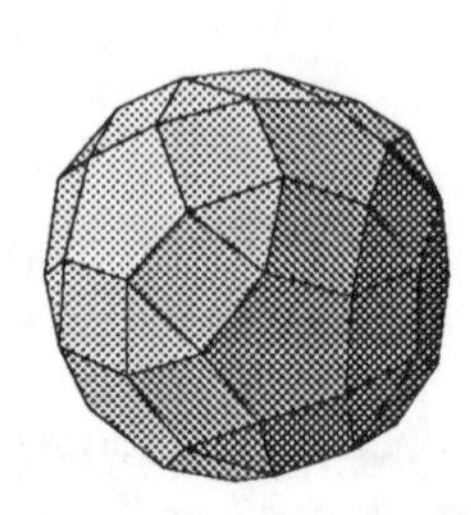

Rhombicosidodecahedron

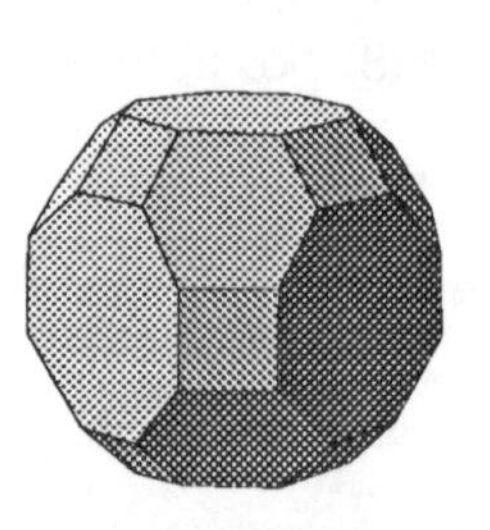

Great rhombicuboctahedron

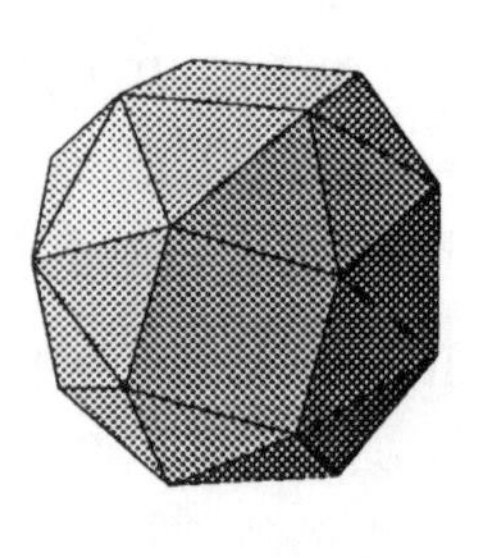

Snub cube

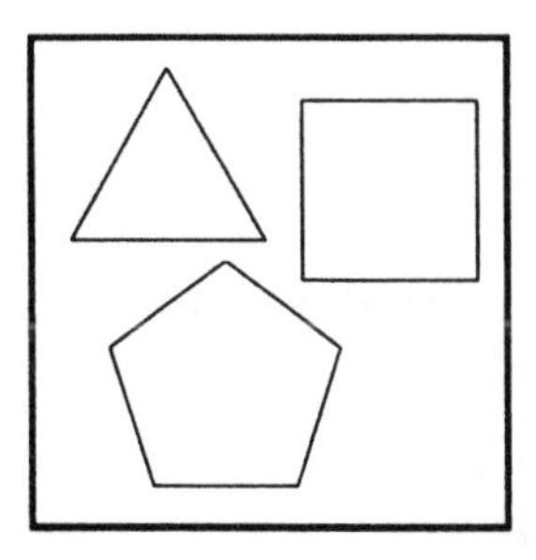

Activity P-1: Regular Polygons

In a **regular polygon,** all sides have the same length and all angles have the same measure.

Here are two equilateral triangles. Although one is larger than the other, their angle measures are the same.

*A triangle whose sides are all equal in length is called an **equilateral triangle**. It is also a regular triangle, for all its angles have equal measure.*

1. What is the measure of each interior angle?

   60°

   Explain how you arrived at your answer.

   See Teacher Comments.

A polygon with more than three sides, such as a rectangle, may have angles that are equal in measure but sides that are not equal in length.

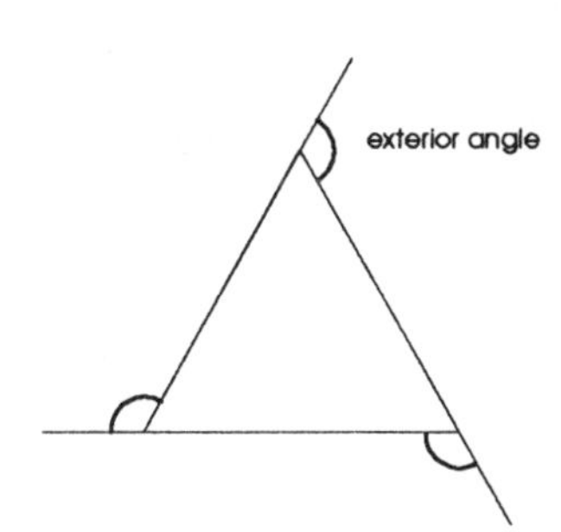

2. What is the measure of each exterior angle of any equilateral triangle?

   120°

   Explain how you arrived at your answer.

   See Teacher Comments.

*A **rhombus** is an example of a polygon whose sides are equal in length but whose angles may not be equal in measure.*

3. What is the sum of the measures of the three exterior angles of any equilateral triangle?

   360°

A regular polygon is both equilateral and equiangular.

Here's a way to check your answer in Exercise 3. Place your pencil flat on the paper as shown in the diagram. Now, rotate your pencil clockwise about its point, turning it through the exterior angle at A, until it lies against side $\overline{AB}$ of the triangle. Slide your pencil along the line through $\overline{AB}$ until the pencil point lies on vertex B. Repeat this procedure—rotate and slide along the side of the triangle—two more times. Your pencil should be back in its original position. Notice that your pencil made a complete turn of 360°.

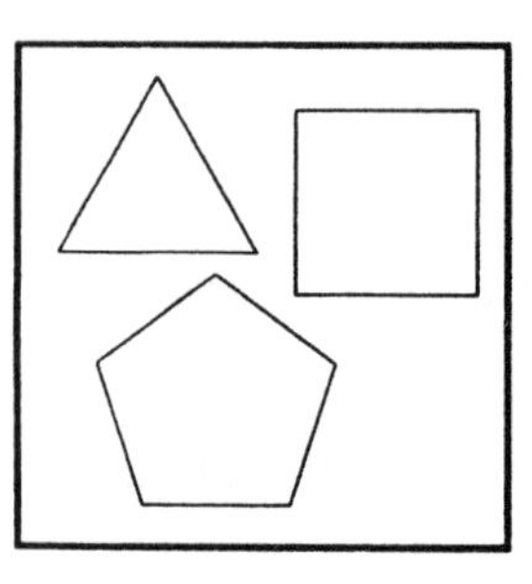

Activity P-1 (continued)

4. Use this pencil method to determine the sum of the exterior angles of the following regular polygons:

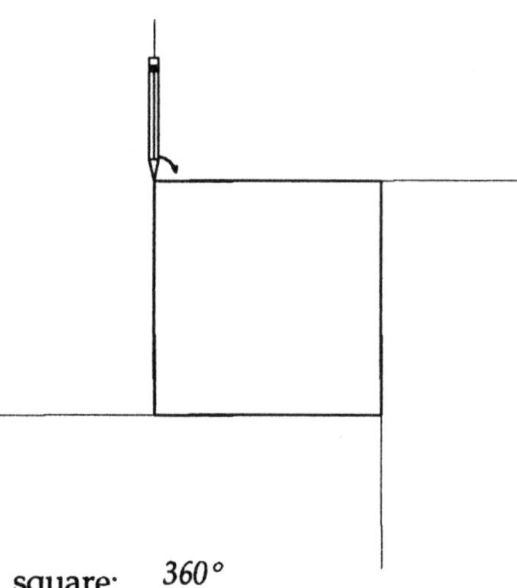

square: *360°*

regular pentagon: *360°*

The conjecture in Exercise 5, when proved, is sometimes known as ***The Exterior Angle Theorem.***

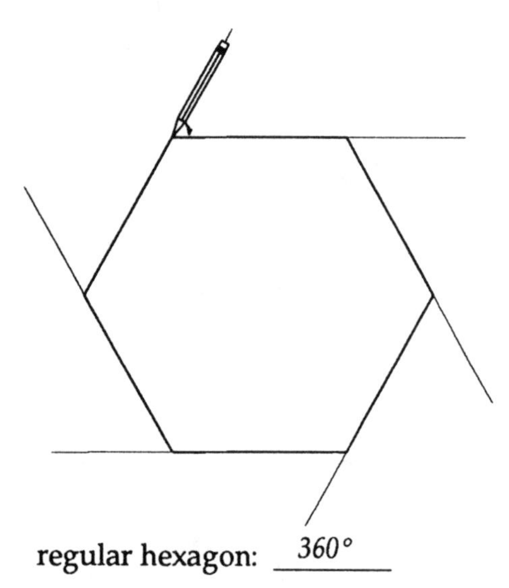

regular hexagon: *360°*

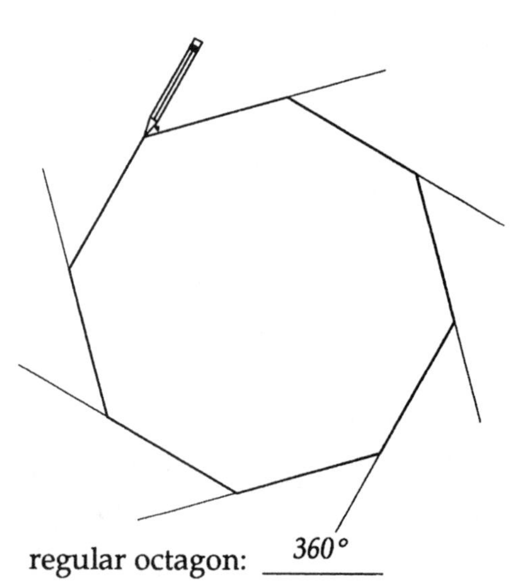

regular octagon: *360°*

5. State a conjecture based on your findings in Exercise 4:

   The sum of the exterior angles of any regular polygon is *360°*.

6. The exterior angles of a regular polygon all have the same measure. Use this fact and your conjecture in Exercise 5 to find the measure of each exterior angle for these polygons:

   square: *90°* regular pentagon: *72°*

   regular hexagon: *60°* regular octagon: *45°*

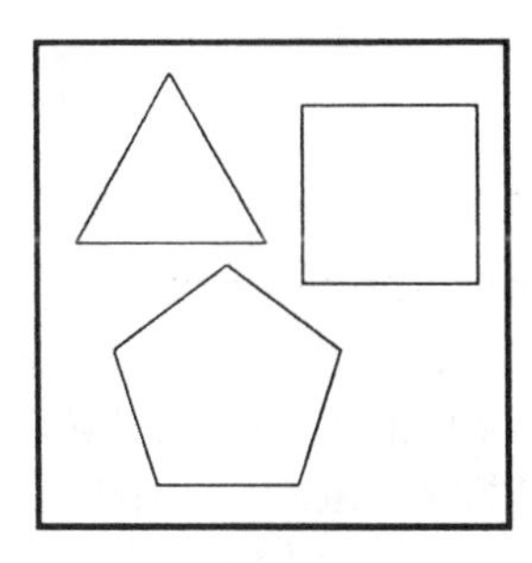

Activity P-1 (continued)

7. Find the measure of each *interior* angle for these polygons:

square: *90°*　　regular pentagon: *108°*

regular hexagon: *120°*　　regular octagon: *135°*

The prefix "tri-" means "three," so the word "triangle" refers to a figure with three angles.

Other polygons are named according to the number of angles they have. The suffix "-gon" means "angle" in Greek and actually comes from the Greek word for "knee." The prefixes "penta-" (5), "hexa-" (6), and "octa-" (8), are common in the English language.

In any language, the most commonly used words are often the most irregular. An example of this is the word "square" for a regular quadrilateral.

8. Put your findings in the chart below:

Name of Polygon	Number of Sides	Each Exterior Angle	Each Interior Angle
Equilateral Triangle	*3*	*120°*	*60°*
Square	*4*	*90°*	*90°*
Regular Pentagon	*5*	*72°*	*108°*
Regular Hexagon	*6*	*60°*	*120°*
Regular Octagon	*8*	*45°*	*135°*

9. Describe a pattern that you see in the chart.

See Teacher Comments.

10. Determine the measure of each interior angle of the following regular polygons:

    a. regular nonagon (9 sides)　*140°*

    b. regular decagon (10 sides)　*144°*

    c. regular dodecagon (12 sides)　*150°*

11. Write an explanation for a friend (who missed class) describing how to find the measure of each interior angle of a regular polygon with *n* sides.

See Teacher Comments.

Teacher Comments for Activity P-1

Before studying the Platonic solids, students must be familiar with the properties of the regular polygons that make up the faces of the polyhedra. This exercise will allow the student to discover ways to easily calculate the interior and exterior angles of polygons. In most textbook treatments, the sum of the interior angles is determined by dissecting the polygon into triangles. We do not use that method here; instead, we find the sum of the exterior angles by rotating a pencil through all the exterior angles.

Background Knowledge: Review the meaning of interior and exterior angle of a polygon. Students should know that a straight angle has a measure of 180° and that the sum of the interior angles of a triangle is 180°.

Presenting the Activity: Before proceeding with the activity, be sure that students understand that a regular polygon must meet both conditions: all sides have the same length, and all angles have the same measure. For a triangle, either condition implies the other; this is not true for polygons with more than three sides. Ask students to produce examples of polygons which fulfill one condition but not the other. (The familiar "house" shape, for example, can be drawn as an equilateral but not equiangular pentagon.)

This activity focuses on the exterior angles of regular polygons, so it is natural to link this with a computer-based activity using turtle graphics. See the polygon programs in H. Abelson and A. diSessa, *Turtle Geometry*, p. 15, or in M. Serra, *Discovering Geometry*, p. 744. Note that to draw a regular polygon using Logo, the turtle turns correspond to the exterior angles. The Exterior Angle Theorem noted in this activity corresponds to the Simple Closed Path Theorem in Abelson and diSessa, p. 24.

Materials: A pencil.

Comments on Activity Questions

1. The three interior angles are equal and have a sum of 180°, so each angle must have a measure of $180°/3 = 60°$.
2. Explanations may vary. Some will say they subtracted 60° from 180° because the straight angle at the vertex is 180°. Others may say that the interior and exterior angles are supplementary.
3. It is standard convention that a polygon with n sides has n exterior angles. Each side is extended in one direction to form an exterior angle at each vertex. Thus, a triangle has three exterior angles.
5. It should be noted that this conjecture is true for any convex polygon, not just the regular ones.
7. Subtract the measure of the corresponding exterior angle from 180°.
9. Many patterns may be observed: as the number of sides increases, the measure of each exterior angle decreases and the measure of each interior angle increases; the sum of the exterior angle and the interior angle is 180°. You might have students develop notation for their patterns. For example, let n be the number of sides, p be the measure of each exterior angle, and q be the measure of each interior angle. Then $p = 360°/n$ and $p + q = 180°$. Graph these equations and compare the graphs. In both cases, as one variable increases, the other decreases. However, only the first case is an example of an inverse variation. The second is a linear function with negative slope.

Teacher Comments for Activity P-1 (continued)

Students may question why the activity doesn't include a 7-sided polygon. Have them calculate the measure of the exterior angle of a 7-sided polygon (360°/7).

10. Emphasize the use of the exterior angle approach used in Exercises 6 and 7.

11. Answers will vary. The method discussed in this activity is to divide 360° by the number of sides to calculate the measure of the exterior angle, and then to subtract that exterior angle measure from 180° to calculate the measure of the interior angle.

Discussion/Extension: How do we reconcile the exterior angle method used here with the usual textbook method? The usual method is to draw diagonals from one vertex of the regular n-gon, producing $(n - 2)$ triangles. Thus, the sum of the interior angles of a regular (or non-regular) n-gon is $180°(n - 2)$. The n interior angles of a regular n-gon are equal, so the measure of each interior angle is $180°(n - 2)/n$. Each exterior angle is supplementary to an interior angle, so the measure of each exterior angle is $180° - (180°(n - 2)/n)$. Show that this is equal to $360°/n$.

The ancient Babylonians used 360° to describe a complete revolution. It probably came from their studies of the earth's orbit around the sun. Discuss what adjustments would have to be made in the preceding activities if they had used 100° to describe a revolution. (While the metric system is quite useful for most measurements, it is not advantageous here.)

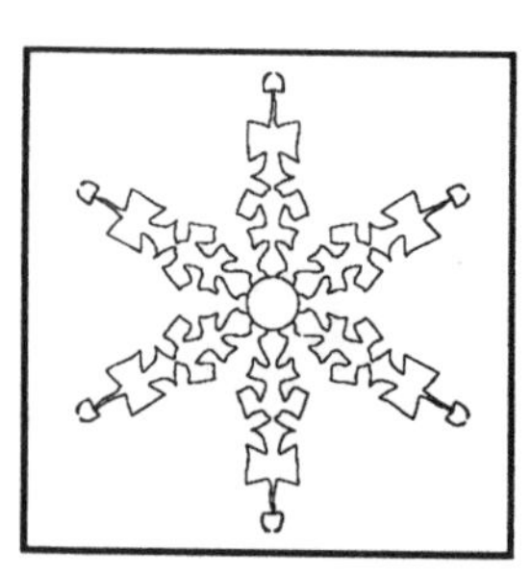

Activity P-2: Symmetry in Two Dimensions

Part I: Rotation Symmetry

The uppercase letter N looks the same right side up or upside down. This letter has **2-fold rotation symmetry**. This means that as you rotate the letter one complete turn (360°) about a point, it appears exactly the same at 2 different positions.

Demonstrate this by placing the tip of your pencil at the center point of the N; now slowly turn the paper halfway around. The N should appear identical to its original self. Make another half turn to return the N to its original position.

N

Rotation symmetry is prevalent in nature. Many microscopic organisms exhibit rotation symmetry. Five-fold symmetry is characteristic of many flowers, while snowflakes provide the best known examples of six-fold symmetry.

Definition: A two-dimensional figure has **rotation symmetry** if it can be turned about a central point (called the **center of rotation**) in such a way that the turned figure appears to be in exactly the same position as the original figure.

Use n to represent the number of times an identical image occurs as the figure is rotated one complete turn (360°). Then we say that the figure has **n-fold rotation symmetry**. (If $n = 1$, we say that the figure has no rotation symmetry.)

1. Consider each of the letters and symbols below:

S A H + × %

a. Which of these figures has no rotation symmetry?

A

b. Which of these figures have 2-fold rotation symmetry?

S H % + X

Two-fold rotation symmetry is often called half-turn or point symmetry.

c. Which have 4-fold rotation symmetry? **+ X**

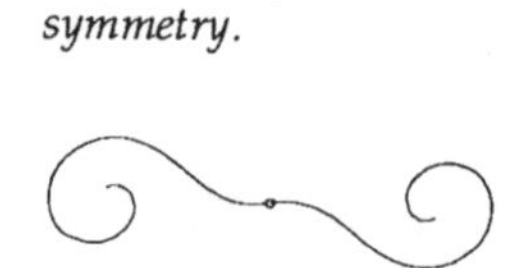

One reason to be interested in regular polygons is that they have many kinds of symmetry. For example, consider the square on the left. Place a square panel over the diagram so that the two squares match exactly. (Ignore the tabs on the square panel.) Place your pencil, eraser side down, in the center hole.

Using your pencil eraser as a center of rotation, turn the square panel clockwise until it again matches up with the diagram.

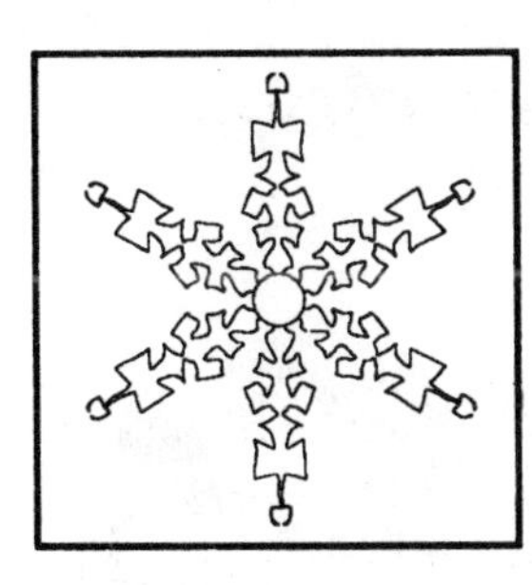

Activity P-2 (continued)

2. a. What fraction of a complete turn did you make?

   1/4

   b. What is the measure of this rotation in degrees?

   90°

   c. A square has _4_-fold rotation symmetry.

3. Use the same technique as in Exercise 2 with a pentagon panel.

   A pentagon has _5_-fold rotation symmetry.

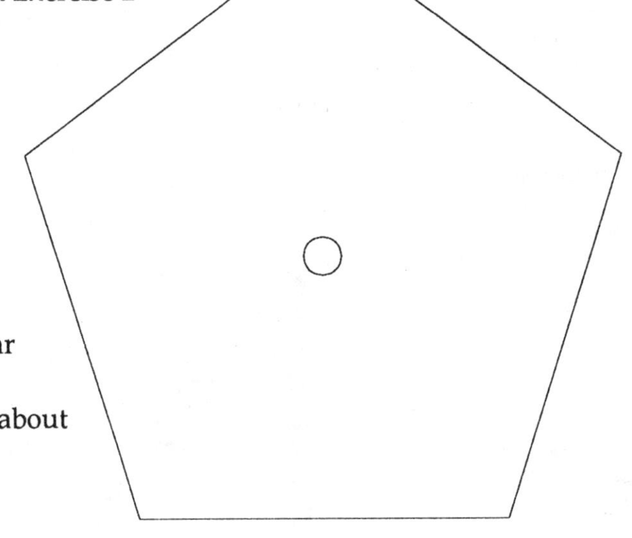

What kind of rotation symmetry does a circle have?

This property makes it possible to turn a round peg smoothly in a round hole.

4. Make a conjecture: A regular polygon with n sides has _n_-fold rotation symmetry about its center.

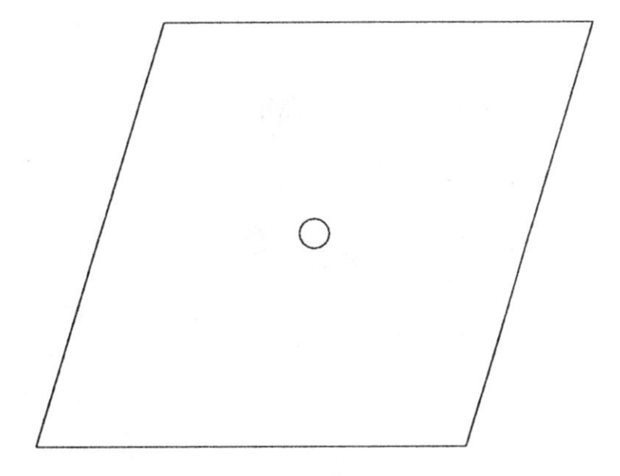

5. Let's try this with a non-regular polygon. Use a rhombus panel.

   a. Why is the rhombus not a regular polygon?

   It is not equiangular.

   b. The rhombus has _2_-fold rotation symmetry.

6. Does your conjecture in Exercise 4 hold for non-regular polygons with n sides?

   No.

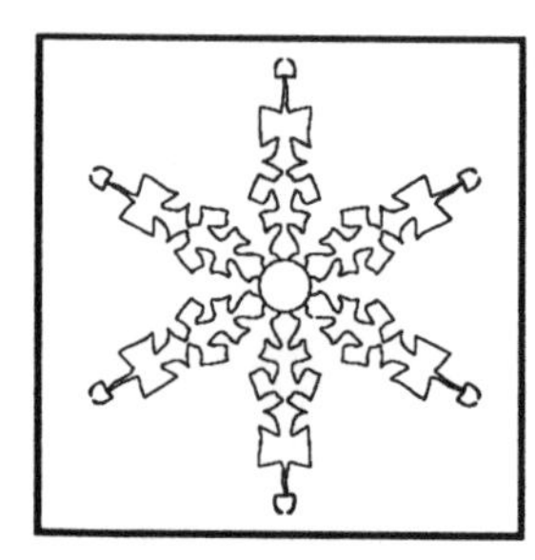

Activity P-2 (continued)

Part II: Reflection Symmetry

The uppercase letter A has **reflection symmetry** about the dotted line shown. To see this, fold your paper along the dotted line. Note that the two halves of the folded A match exactly. The fold in your paper demonstrates an **axis of reflection** for the A.

Reflection symmetry is sometimes called ***line symmetry****, or* ***bilateral symmetry****. Most animals, including people, exhibit bilateral symmetry, at least externally. Notice that the right side of your body is pretty much a mirror image of the left side.*

Definition: A two-dimensional figure has **reflection symmetry** about a line (called the **axis of reflection**) if the pattern on one side of the line is the mirror image of the pattern on the other side.

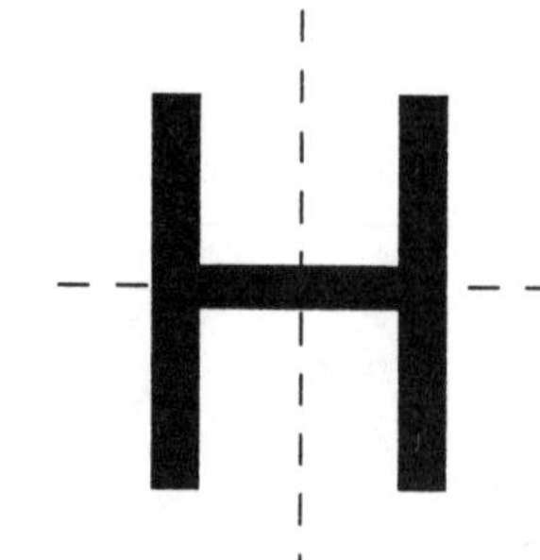

7. Find two axes of reflection for the H in the diagram. Sketch these lines on the picture.

The axis of reflection is sometimes called a ***mirror line****.*

8. Sketch all axes of reflection, if any, for each of the letters and symbols below.

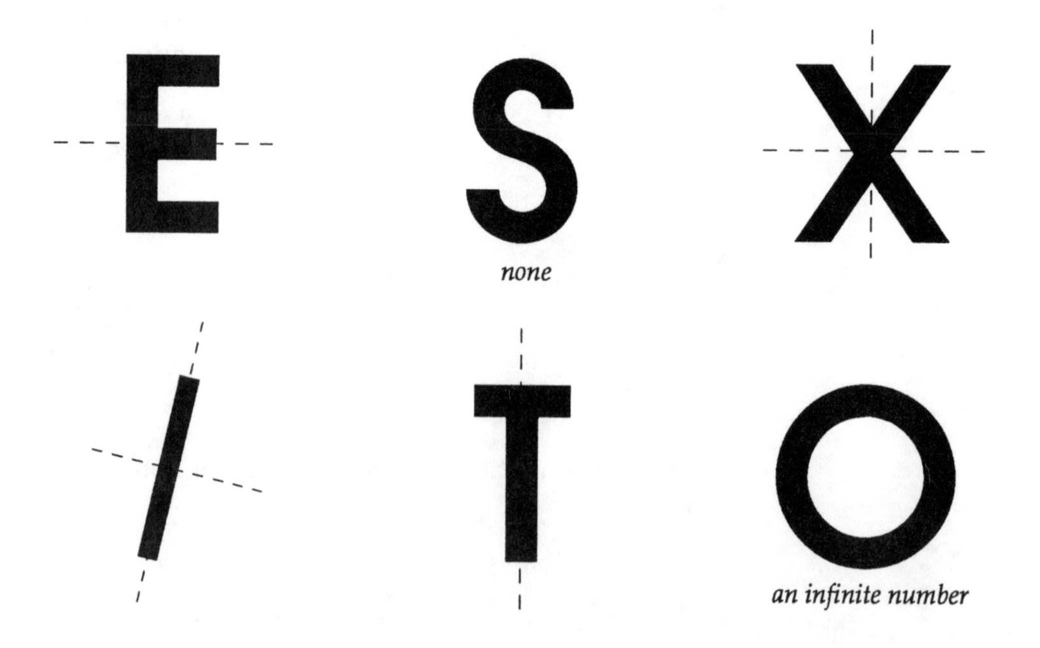

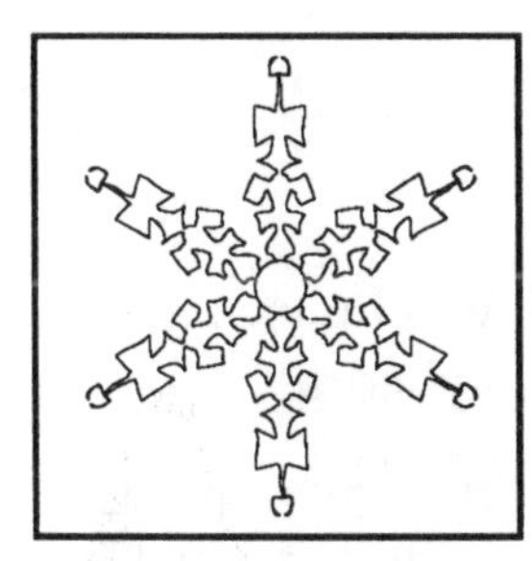

Activity P-2 (continued)

Regular polygons have several axes of reflection. In the following exercise you will make a square, complete with its axes of reflection.

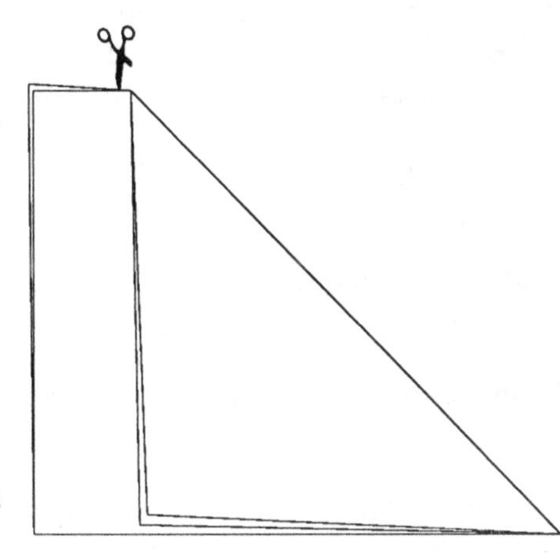

9. Take any rectangular piece of paper and, by matching opposite edges, fold it carefully into quarters. Hold it so that the center of the original paper is at the bottom right. Now fold it one more time so that you bisect the 90° angle at the bottom right. Cut off the excess paper so edges match.

   Now open up your paper. You have a square in which each fold lies on an axis of reflection.

   a. How many axes of reflection does your square have?

      Four; two are diagonals and two are side bisectors.

   b. Is this true for any square?

      Yes.

A mirror kaleidoscope can be used to form the image of a regular polygon from just one line segment.

Other regular polygons can be seen by changing the angle between the mirrors.

10. Find as many axes of reflection as you can for a triangle, pentagon, and rhombus panel. Draw them with ballpoint pen on your polygon panels and fold along the lines to see if you are right. Then sketch the reflection lines on the shapes below.

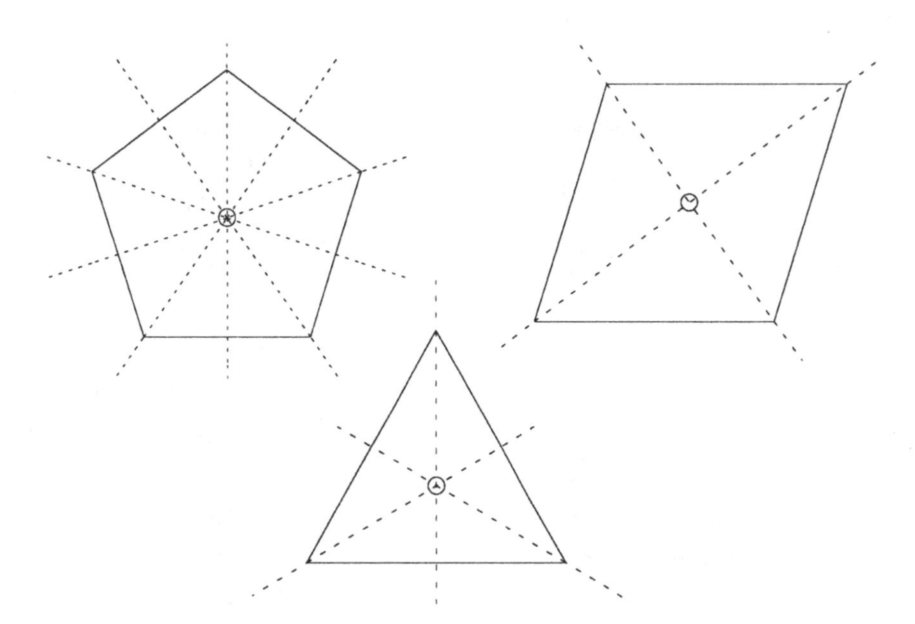

11. Make a conjecture: A regular polygon with n sides has $\underline{n}$ reflection axes.

Teacher Comments for Activity P-2

In this activity, students will learn to identify rotation and reflection symmetries of two-dimensional figures. This exercise will prepare them to work with the symmetry properties of polyhedra.

Background Knowledge: Students should know the following definitions: regular polygon, square, pentagon, and rhombus. (A rhombus is a quadrilateral with four equal sides; the angles are not required to have equal measure. A square is a rhombus with equal angles.)

Presenting the Activity: Emphasize the difference between regular and non-regular polygons; their high degree of symmetry is one thing that distinguishes regular polygons.

Materials: One triangle, square, pentagon and rhombus panel; an extra rectangular sheet of paper. Optional: one or two small rectangular mirrors.

Comments on Activity Questions

Part I. Here is a more precise definition of rotation symmetry: a figure has n-fold rotation symmetry if a copy turned through an angle of $360°/n$ coincides with the original. Note that symbols with 4-fold rotation symmetry also have 2-fold symmetry, since two turns of 90° equal a turn of 180°. You can demonstrate rotation symmetry using an overhead projector. Make two transparencies of the same figure, overlay them so they match, and write TOP on the upper one. Turn the top transparency so that its figure rotates onto the underlying one; the word TOP will be seen in rotated position.

Part II. Here is a more precise definition of reflection symmetry: a figure has reflection symmetry with respect to a line m if a copy of the figure reflected in m coincides with the original. Each point P on the figure has a reflection image P' on the figure and m is the perpendicular bisector of the segment PP'. You can demonstrate reflection symmetry using an overhead projector. Overlay and match two transparencies of the same figure showing a line of reflection symmetry, and write TOP on the upper one. Turn over the top transparency so that the figure and reflection axis coincide with the underlying ones; the word TOP will be seen in mirror image.

7. Note that the **H** has a center of 2-fold rotation where the reflection axes intersect.

8. Mirror Test for reflection symmetry: place a mirror along the reflection axis of the figure, perpendicular to the paper. Half of the figure shows on the paper, while the other half appears as a reflection in the mirror.

9. Encourage students to describe the reflection axes of the square: two are diagonals, the other two bisect the sides of the square.

10. Compare the reflection axes of the rhombus with those of the square: only the diagonals of the rhombus are reflection axes, not the side bisectors. Ask students to find the two reflection axes of a non-square rectangle: only the side bisectors are reflection axes, not the diagonals.

11. Regular polygons have n reflection axes that cut through the n vertices of the polygon and bisect the n sides of the polygon. All of the reflection axes intersect at the center of the polygon, which is a center of n-fold rotation. Such figures are said to have n-fold kaleidoscopic symmetry—this is the symmetry of designs seen in a kaleidoscope. The whole design in a kaleidoscope is created by one small piece of the design that is reflected repeatedly in two mirrors that meet at an angle of $180°/n$.

Teacher Comments for Activity P-2 (continued)

Use two mirrors, standing upright, to form the image of a regular polygon from one line segment drawn on paper. Change the angle between the mirrors to form the image of a different regular polygon. MIRA activities can be incorporated into this lesson, as well.

Discussion/Extension: Some students may want to show their ability to write "backwards." Usually this "mirror-writing" can be read only by holding a mirror parallel to the plane. They might also want to try writing a word that looks the same when reflected in a mirror. MOM, TOOT, or any other palindrome where each letter has a vertical reflection axis will work. For an example with horizontal reflection symmetry, use the word CHOICE. Some words and even sentences have half-turn, or "upside down" symmetry: NOW NO SWIMS ON MON. See Martin Gardner, *The New Ambidextrous Universe*, for a good discussion. *Inversions*, by Scott Kim, has many examples of words with reflection or rotation symmetry; his accompanying interactive computer program *Letterforms and Illusion* encourages students to design their own symmetric words and forms.

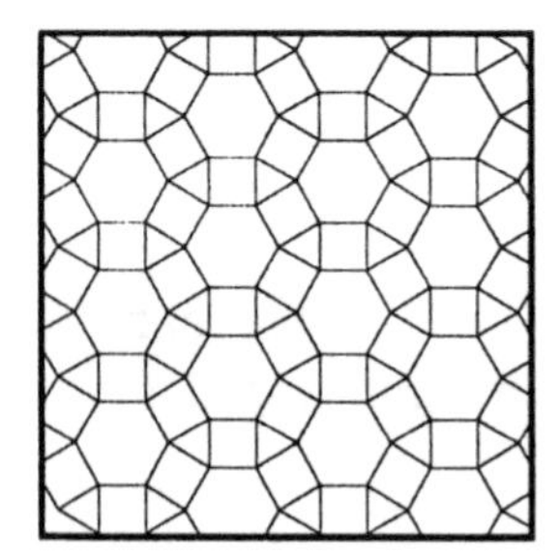

Activity P-3: Polygon Tessellations

Arrange several of the square panels as if tiling a floor. Make sure that the panels do not overlap and that the covering has no gaps, except for the holes in the centers and at the corners of the panels. Notice that this pattern of square panels could be extended indefinitely.

Examples of tilings can be found in floor patterns and on wallpaper. Beautiful mosaic tilings can be seen in Moorish mosques and palaces, such as the Alhambra in Granada, Spain.

Definition: An arrangement of closed shapes which covers the plane without gaps or overlaps is called a **tessellation** or **tiling**.

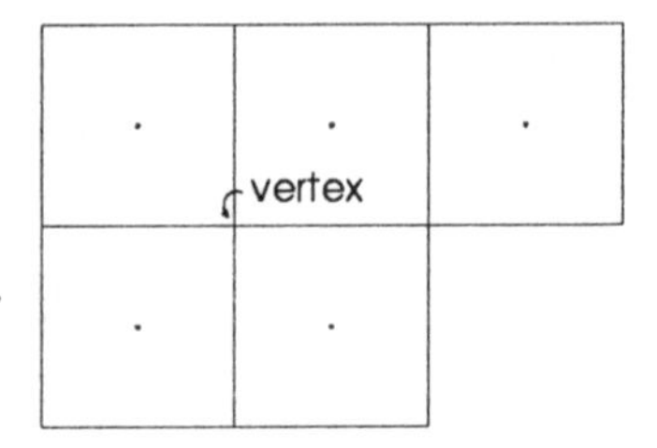

1. Put your finger on one vertex in your tessellation which is completely surrounded by squares.

   a. How many squares surround your finger?

      Four.

   b. How many angles share that vertex?

      Four.

   c. What is the total angle measure around that vertex?

      360°

There are several ways to form tessellations using polygons. The one you have made with square panels is an example of a **regular tessellation**, a tessellation using only one regular polygon shape.

The Dutch artist M.C. Escher (1898-1972) was fascinated by tilings. His brilliant idea was to form repeated patterns of recognizable figures, such as birds, fish, or reptiles. Below is an Escher-like tiling.

2. Experiment with other regular polygon panels. What other regular tessellations can you make?

   Triangular and hexagonal tessellations.

3. Explain why you cannot make a regular tessellation using only regular octagons.

   The measure of each interior angle of a regular octagon is 135°, which is not a factor of 360°.

4. Can you make a tessellation using only the rhombus panels?

   Yes. See Teacher Comments.

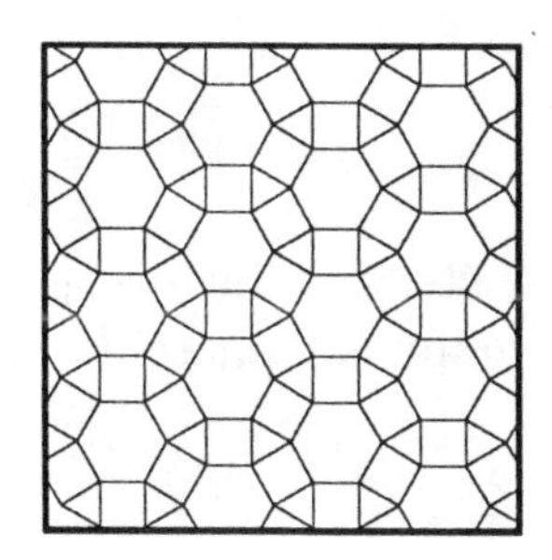

Activity P-3 (continued)

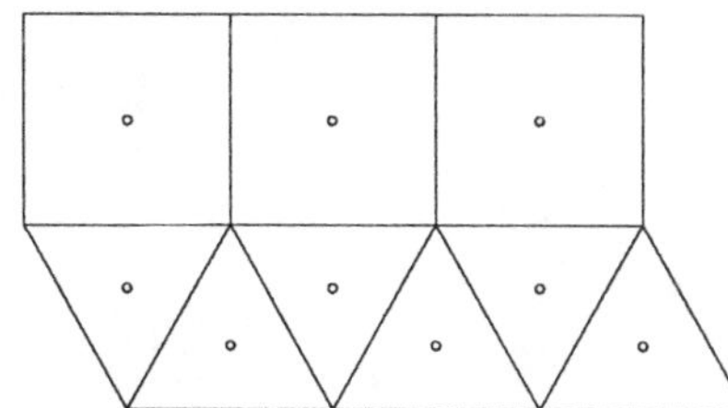

A tessellation that uses more than one type of regular polygon as its tiles is called a **semiregular tessellation** if the arrangement of tiles at each vertex is the same. Use triangles and squares to make the semiregular tessellation shown in the diagram.

Regular pentagons don't tile, but many equilateral (though not equiangular) pentagons do.

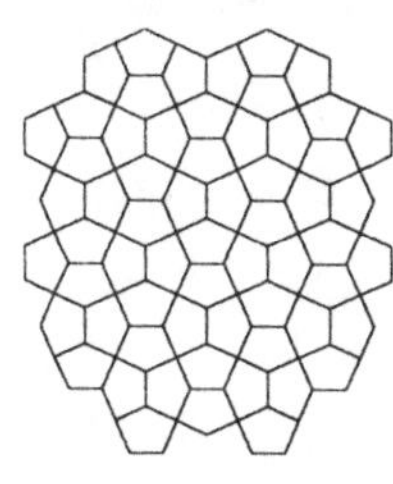

This pattern is seen in street tiling in Cairo and in the mosaics of Moorish buildings. A similar tiling can be obtained as the dual of a semi-regular tiling (see exercise 8).

5. Put your finger on one vertex in your tessellation of squares and triangles.

   a. Describe the arrangement of the shapes surrounding your finger.

      Three triangles followed by two squares.

   b. How many angles surround that vertex?

      Five.

   c. What is the total angle measure around that vertex?

      $3(60°) + 2(90°) = 360°$

   d. Choose another vertex in the tessellation. Is the arrangement of the tiles surrounding that vertex the same as in 5a?

      Yes.

6. a. Find a different arrangement of triangles and squares that forms a semiregular tessellation. Sketch part of this tessellation. Describe the order of the shapes that surround each vertex.

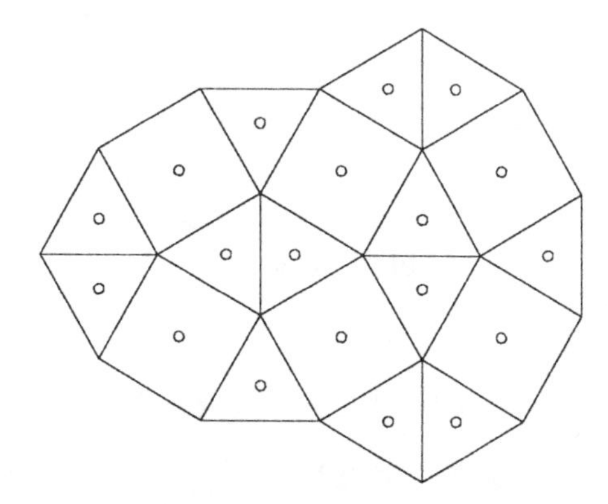

Two triangles followed by square, triangle, square.

   b. Experiment with the regular polygon panels to find other examples of semiregular tessellations. Make a sketch on the back of this sheet of each one that you find. *See Teacher Comments.*

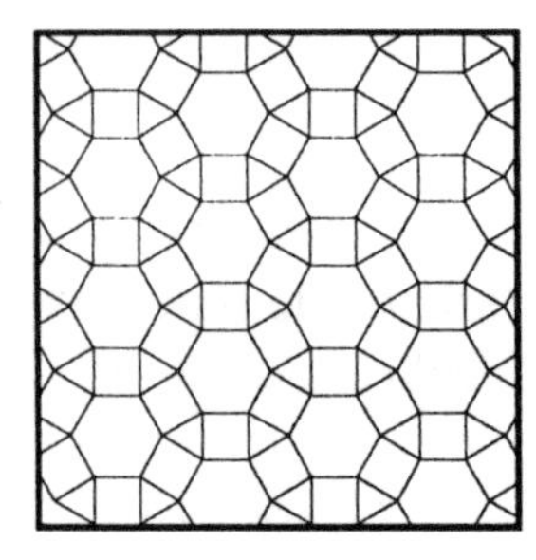

Activity P-3 (continued)

Each tessellation using regular polygons has a companion tessellation which is called its **dual**. The dual tessellation is drawn by connecting centers of adjacent tiles in the original tessellation.

7. This diagram shows a regular tessellation of equilateral triangles, with the center of each triangle marked. Use a straightedge and pencil to connect the centers of two adjacent triangles. Continue to do this for the entire tessellation. The new tessellation you have made is the **dual** of the original tessellation.

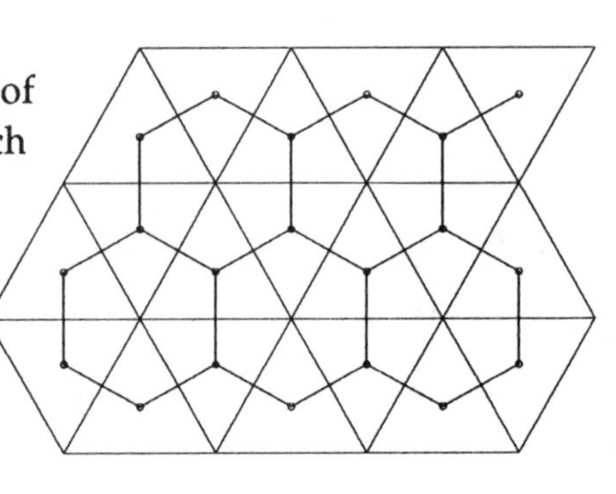

When drawing the dual of a tessellation, the line segment connecting the centers of two adjacent shapes must cross the common edge of the two shapes.

What shapes are the tiles in the dual tessellation?

Hexagons.

8. Draw the dual of the semiregular tessellation shown by connecting the centers of adjacent polygons.

The dual of a regular tessellation will be a regular tessellation. What about the dual of a semiregular tessellation?

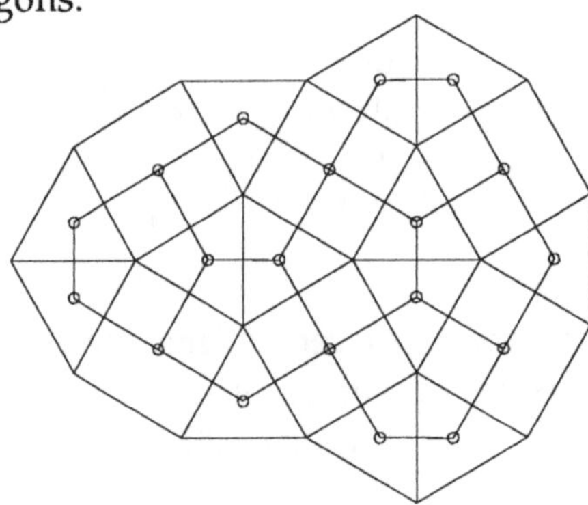

a. Is the dual tessellation a regular tessellation?

No.

b. Is it a semiregular tessellation?

No.

c. What shapes are the tiles in the dual tessellation?

Non-regular pentagons with four equal sides and two right angles.

d. How many tiles are at each vertex of the original semiregular tessellation?

Five.

Teacher Comments for Activity P-3

This activity deals with regular, semiregular, and dual tessellations. Each of these topics has an analogous three-dimensional component: regular tessellations can be compared with Platonic solids, semiregular tessellations with Archimedean solids, and dual tessellations with dual polyhedra. In addition, the problem of packing solids to fill space is related to tiling by polygons to fill the plane.

Background Knowledge: Regular polygons and their interior angle measures (Activity P-1).

Presenting the Activity: Students may want to make a list of the angle measures of the regular polygons or write the measures directly on the angles of the polygon panels. Arranging students in pairs or groups of four will let them combine their panels to discover new patterns.

Materials: Several of each kind of polygon panel: at least 12 triangles, 9 squares, 9 rhombuses, and 7 hexagons. Optional: a container of colorful tiles of regular polygons, such as *Pattern Blocks*.

Comments on Activity Questions

2. A beehive demonstrates the regular hexagonal tessellation. See H. Steinhaus, *Mathematical Snapshots*, p. 75.

4. This exercise shows how a parallelogram can be used to make a tessellation. In fact, any quadrilateral will tessellate the plane. See, for example, D. Seymour and J. Britton, *Introduction to Tessellations*, H. M. S. Coxeter, *Introduction to Geometry*, or P. O'Daffer and S. Clemens, *Geometry: An Investigative Approach*, Chapter 3.

5a. It's conventional to name a polygon with the fewest number of sides first and proceed clockwise to name the others that surround that vertex. One effective way to describe semiregular and regular tessellations is the Schläfli symbol. This code tells the type of polygons and the (cyclic) order in which they occur at any vertex, e.g., 33344 (or $3^3 4^2$) denotes three triangles followed by two squares.

5d. The arrangement of the tiles at each vertex must be identical in a semiregular tessellation. There are highly symmetric tessellations with two or more different vertex configurations of polygons; they are called demiregular, or more precisely, *k*-uniform tessellations. See B. Grünbaum and G. C. Shephard, *Tilings and Patterns*.

6a. Make sure that students have the same arrangement of polygons at each vertex (same vertex configuration).

6b. Five semiregular tessellations using combinations of triangles, squares, or hexagons are possible. Exactly eight semiregular tessellations exist altogether. See Student Project 1 for an analysis of the semiregular tessellations. Other references include P. O'Daffer and S. Clemens, *Geometry: An Investigative Approach*, D. Seymour and J. Britton, *Introduction to Tessellations*, K. Critchlow, *Order in Space*, H. M. Cundy and A. Rollet, *Mathematical Models* (Section 2.9), and B. Grünbaum and G. C. Shephard, *Tilings and Patterns*.

7. In geometry, the concept of duality involves interchanging vertices and faces. In the case of a dual tessellation, the center of each face of the original tessellation becomes a vertex of the dual. Each vertex configuration of the original determines the shape of a face of the dual. Duality in three dimensions, the subject of Activity 6, is an analogous concept.

   Note that the dual tessellation of a regular tessellation is a regular tessellation, and that the square tessellation is self-dual.

Teacher Comments for Activity P-3 (continued)

8. The duals of semiregular tessellations are not regular or semiregular, but do consist of congruent shapes, since a semiregular tessellation has exactly the same arrangement of tiles at each vertex.

   Although the pentagons in the dual tessellation here are not regular, they are highly symmetric. You might have students draw the dual of the semiregular tessellation in Exercise 5 and compare it to this one.

Discussion/Extension: Have students discuss various uses for these patterns. Some may be appropriate for wallpaper, floor tiles, fabric or wrapping paper. Several books explain how to make Escher-like tessellations. Some examples are M. Serra, *Discovering Geometry*, and D. Seymour and J. Britton, *Introduction to Tessellations*. Students may like the challenge of creating a tessellation that will wrap around a particular box while maintaining the continuity of the pattern. D. Schattschneider and W. Walker have done just that in *M. C. Escher Kaleidocycles*.

The problem of tiling the plane with convex non-regular pentagons is not completely solved. For more on tiling with pentagons and hexagons, see M. Gardner, *Time Travel and Other Mathematical Bewilderments*, Chapter 13, and D. Schattschneider, "Tiling the Plane with Congruent Pentagons."

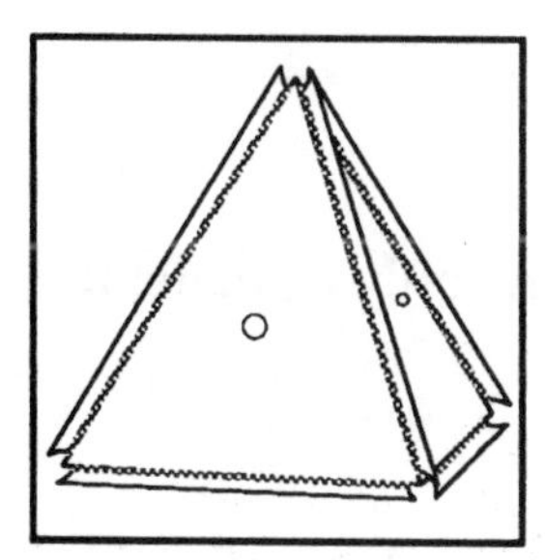

Activity 1: The Regular Polyhedra

*A polyhedron is a three-dimensional shape formed by joining edges of polygons to enclose a region of space. The polygons are called the **faces** of the polyhedron. Exactly two polygons meet at each **edge** of the polyhedron. At least three faces meet at each **vertex** of the polyhedron.*

1. Using some of the polygon panels, make a polyhedron in the following way:
   - Connect the edges of two polygons by holding the tabs together with one hand and fastening a rubber band over them with the other hand.
   - Continue to connect polygons together until you have completely enclosed some region of space.
   - You must be sure that there are at least three faces at each vertex and exactly two faces at each edge of the polyhedron.

Describe your polyhedron: *Answers will vary.*

a. How many faces are there?

b. Which polygons did you use for the faces?

c. What configurations of polygons do you have at the vertices?

*Two polygons are **congruent** if their corresponding sides are equal in length and their corresponding angles are equal in measure. Two congruent polygons have the same size and shape.*

Definition: A **regular polyhedron**, or Platonic solid, is a polyhedron with the following properties:

a. All faces are regular polygons.

b. All faces are congruent to each other.

c. The same number of faces meet at each vertex in exactly the same way.

2. Examine the polyhedron you made in Exercise 1. Is it a regular polyhedron? If it is not a regular polyhedron, explain which of the necessary properties is (are) missing.

   Answers will vary.

3. Build as many different regular polyhedra as you can using squares only. Remember that you must have the same number of faces at each vertex. How many can you find?

   One.

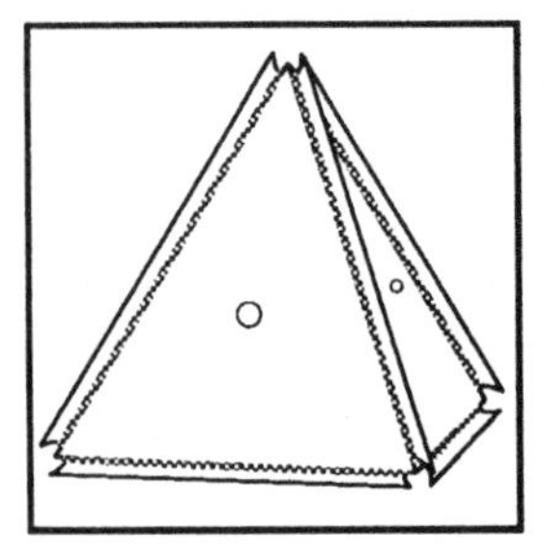

Activity 1 (continued)

4. Build as many different regular polyhedra as you can using only equilateral triangles for faces. How many can you find?

   Three.

5. Build as many different regular polyhedra as you can using only regular pentagons for faces. How many can you find?

   One.

Dice made in the shape of regular polyhedra are used in games such as Dungeons and Dragons™ because each face has an equal likelihood of landing flat. The cube is the most common shape for dice. In fact, the word for dice in Greek is 'cubos.'

6. Why is it impossible for a regular polyhedron to have faces which are regular hexagons? (Remember that at least three faces must meet at each vertex.)

   The measure of each interior angle of a regular hexagon is 120°. Three such faces at one vertex will form an angle sum of 360°, causing the faces to lie flat.

7. Why is it impossible for a regular polyhedron to have faces with more than six sides?

   A face with more than six sides will have an interior angle measuring more than 120°. Three such faces will form an angle sum greater than 360°.

8. Your answers in Exercises 3-7 should convince you that there are exactly five regular polyhedra. Their names describe how many faces they have.

   a. For each polyhedron, count the number of faces it has and enter this number in the first column in the chart below.

Here are some other polyhedra names that are based on the number of faces of the polyhedron:

pentahedron—5 faces
decahedron—10 faces
tetrakaidecahedron—14
pentakaidecahedron—15
hexakaidecahedron—16

What do you suppose the word "kai" means?

Number of Faces	Name	Kind of Faces	Number of Faces at Each Vertex
6	*Hexahedron*	*Squares*	3
4	*Tetrahedron*	*Triangles*	3
8	*Octahedron*	*Triangles*	4
20	*Icosahedron*	*Triangles*	5
12	*Dodecahedron*	*Pentagons*	3

   b. From the following list, choose the appropriate name for each shape. The prefix of each name tells the number of faces a polyhedron has. (You may want to use a dictionary to help you.) Enter the names in the chart.

   Names: dodecahedron, tetrahedron, hexahedron, octahedron, icosahedron

   c. To complete the chart, note the kind of face (triangle, for example) that the polyhedron has and the number of faces that meet at each vertex.

Teacher Comments for Activity 1

In this activity, students will construct polyhedra by connecting panels from the kit with rubber bands. They will discover that there are exactly five regular polyhedra, which are known as the Platonic solids. This demonstration is shown in the animated videotapes, *The Platonic Solids* (Visual Geometry Project) and *Regular Convex Polyhedra: Why Exactly Five?* (Cal. State, Northridge).

Background Knowledge: Congruence; regular polygons and the measures of their angles.

Presenting the Activity: You will probably want students to do this activity before viewing the videotape. This activity is very effective when done in a group.

Materials: Polygon panels (triangles, squares, pentagons, hexagons, rhombuses) and rubber bands (size 10). To make all five Platonic solids, each group of students needs 32 triangles, 6 squares, and 12 pentagons. Extra panels, including hexagons and rhombuses, are useful for experimentation.

Comments on Activity Questions

Many textbooks require that a regular polyhedron be convex. We do not list convexity as a requirement, but note that it is a consequence of our three conditions a, b, and c. Condition c can be sharpened to say that all dihedral angles (angles between faces) are the same. It is possible to have the same number of faces meet at each vertex, but not "in the same way": just invert one dome of five faces on the icosahedron. The polyhedron that results is not regular.

3. Four squares at a vertex will make a flat tessellation rather than a polyhedron.

4. There are several nonregular polyhedra with equilateral triangle faces. Convex polyhedra with equilateral triangle faces are called deltahedra; these are analyzed in Student Project 8. An example of a non-convex polyhedron with equilateral triangle faces is the stella octangula. See Visual Geometry Project, *The Stella Octangula*.

6. Three hexagons at a vertex will make a flat tessellation rather than a polyhedron.

7. As the number of sides of a polygon increases, the measure of each interior angle increases (see Activity P-1). So in a polygon with more than six sides, each angle must be greater than 120°. Three or more such polygons cannot meet at a single vertex without overlapping or crumpling.

Discussion/Extension: Why are there only five regular polyhedra? Here's an algebraic explanation:

Let p be the number of sides on one face of the polyhedron. Then each interior angle of the face has measure $180° - 360°/p$ (see Activity P-1).

Let q be the number of faces that meet at each vertex. In order to form a polyhedron, we must require $q(180° - 360°/p) < 360°$. This is equivalent to the inequality $(p-2)(q-2) < 4$.

Since p and q must be integers greater than or equal to three, the only solutions to this inequality are $(p, q) = (3, 3), (3, 4), (3, 5), (4, 3)$, or $(5, 3)$.

This algebraic explanation proves only that no more than five regular polyhedra exist. To prove that exactly five exist, we would still have to build them all. Note that each of the solutions (p, q) does correspond to a Platonic solid: (3, 3)⇔tetrahedron, (3, 4)⇔octahedron, (3, 5)⇔icosahedron, (4, 3)⇔cube, and (5, 3)⇔dodecahedron.

A similar algebraic analysis is used in Student Project 8, Deltahedra. In this case, the algebraic equation has 19 possible solutions, but only 8 deltahedra can actually be built.

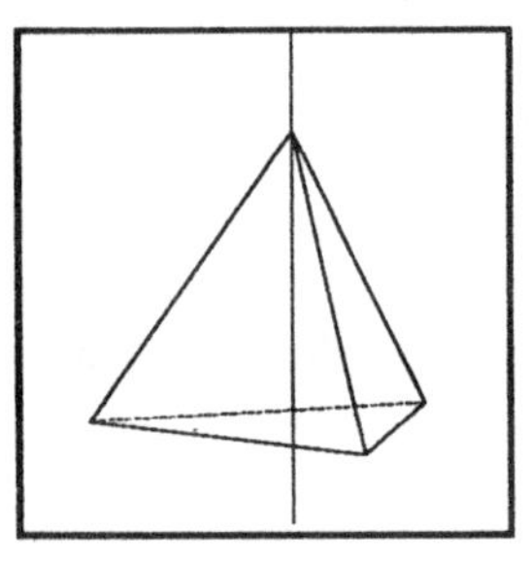

Activity 2: Symmetry of Regular Polyhedra

The five regular polyhedra are known as Platonic solids, after the Greek philosopher Plato. Plato was so taken with the beauty and regularity of the solids that he used them to represent the elements thought to make up the world.

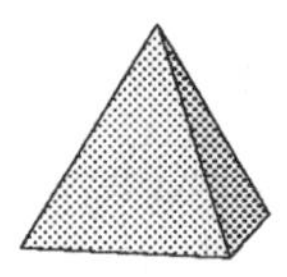 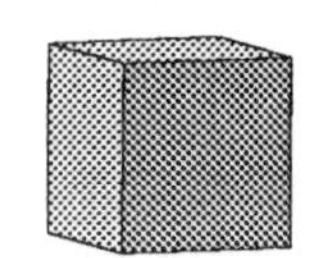 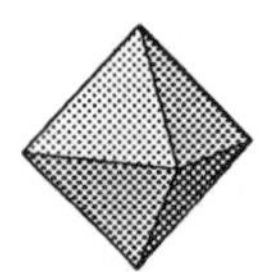 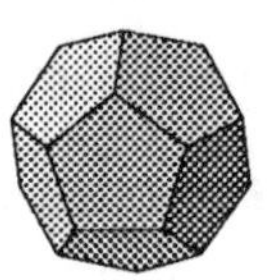 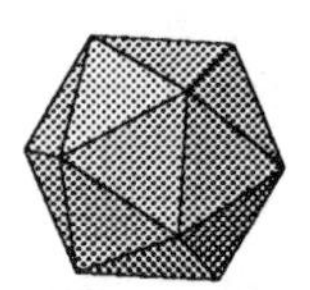

Plato (b. 429 B.C.) was a Greek philosopher. Even though the regular polyhedra were known well before his time, they are referred to as ***Platonic solids*** *because he wrote about them in his dialogue* Timaeus. *In this allegory, the polyhedra represent the elements of the physical world.*

1. Look at your models of the five Platonic solids. What would you say is "regular" (uniform) about each one?

   They look the same when viewed at any corner, edge, or center of any face.

   They have congruent regular polygon faces.

   They have congruent edges, face angles, and corners.

One property that makes the shapes of the Platonic solids so pleasing is their symmetry. In the exercises below, you will investigate the symmetries of some of the Platonic solids.

Rotation Symmetry

Examples of Platonic solids abound in nature. Crystals of pyrite are often shaped like cubes; alum crystals occur as octahedra. The atoms of a methane molecule are arranged like the corners of a tetrahedron.

Many viruses are shaped like icosahedra, while the skeletons of some miscroscopic animals called radiolaria *are shaped like dodecahedra, octahedra, or icosahedra.*

A three-dimensional figure is turned about an **axis of rotation** to determine its rotation symmetry.

2. Pick up your regular octahedron and put a pipe cleaner or straw through a pair of opposite vertices. Hold the straw vertically and look directly down on it; now turn the octahedron a quarter turn. You should see that the octahedron looks as if you had not turned it—faces, vertices, and edges appear to be in the same position as before. The straw demonstrates an axis of **4-fold rotation symmetry**.

   How many different axes of 4-fold rotation symmetry does the octahedron have? (Hint: How many pairs of vertices are there?)

   Three.

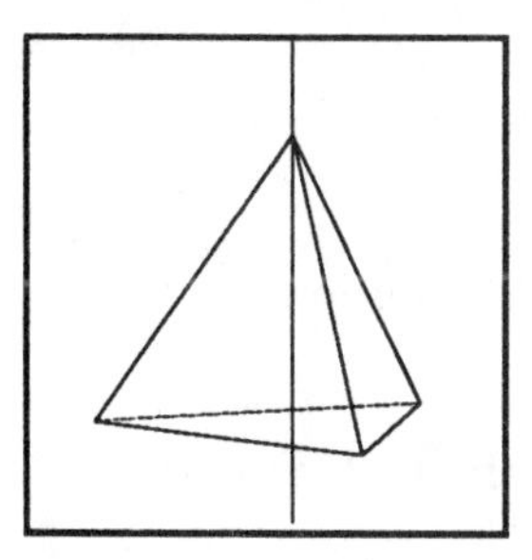

Activity 2 (continued)

3. The cube also has axes of 4-fold rotation symmetry, but they do not go through the vertices. Holding your cube, put a pipe cleaner or straw through the centers of two opposite faces. Turn the cube around this axis 1/4 of a turn; it looks just like it did in its original position.

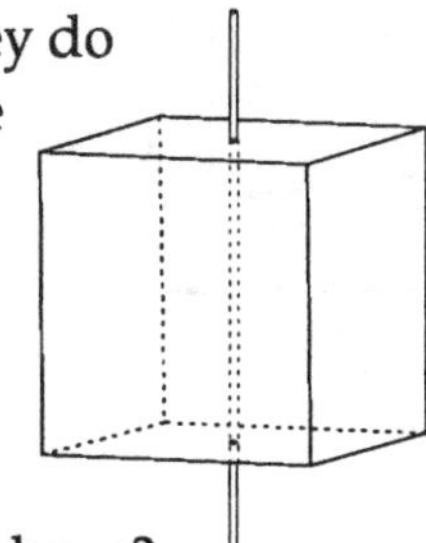

   How many axes of 4-fold rotation symmetry does the cube have?

   Three.

4. The cube has axes of 3-fold rotation symmetry as well. Use a pipe cleaner or straw to connect diagonally opposite corners of the cube, piercing through the center of the cube. Hold the axis (the straw) vertically and look directly down on the model. Slowly turn the axis until the faces, vertices, and edges of the cube first appear to be in the same position as before.

Why are the symmetries of the cube and octahedron alike? It is because they are ***dual*** *polyhedra; the vertices of one correspond to the faces of the other, and vice versa. The dual relationship of a pair of objects often serves as a shortcut, allowing us to determine the properties of both at the same time.*

   a. How far have you turned?

      120°, or 1/3 of a complete turn.

   b. How many axes of 3-fold rotation symmetry does the cube have?

      Four.

5. Use a pipe cleaner or straw to demonstrate an axis of 3-fold rotation symmetry on your regular octahedron.

   a. Where is it located?

      It passes through the centers of a pair of opposite faces.

   b. How many such axes of 3-fold rotation symmetry does the octahedron have?

      Four.

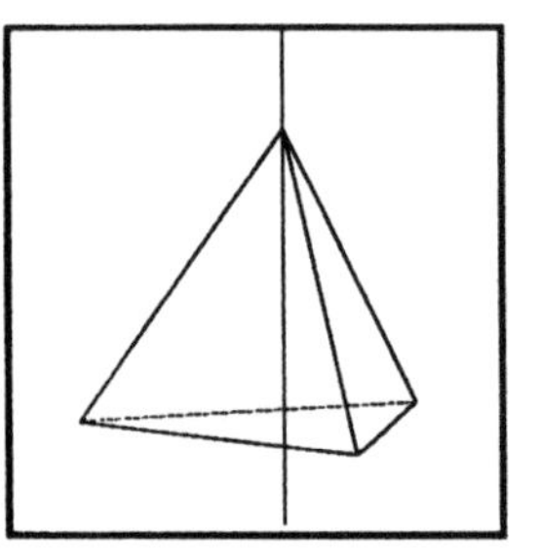

Activity 2 (continued)

6. Use a pipe cleaner or straw to demonstrate an axis of 3-fold rotation symmetry on your regular tetrahedron.

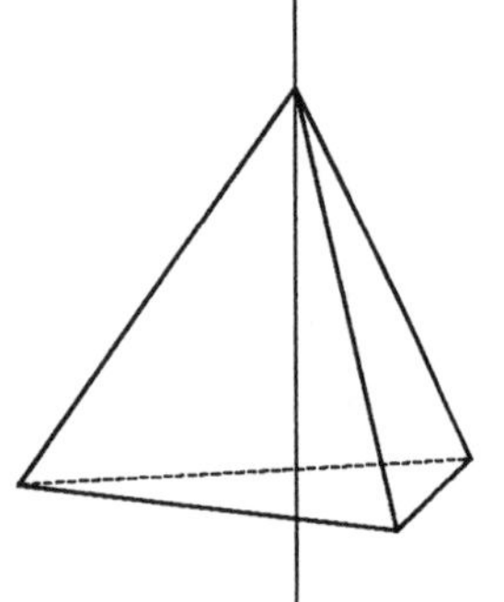

   a. Where is it located? Draw it on the picture shown here.

      It passes through a vertex and the center of the opposite face.

   b. How many such axes of 3-fold rotation symmetry does the tetrahedron have?

      Four.

The tetrahedron does not have opposite vertices or opposite faces, so a regular tetrahedron has only two kinds of rotation symmetry.

7. The cube, the octahedron, and the tetrahedron have another kind of rotation symmetry, in which the axis of rotation connects the midpoints of two opposite edges and passes through the center of the figure. Use a straw to demonstrate one such axis of rotation on each polyhedron.

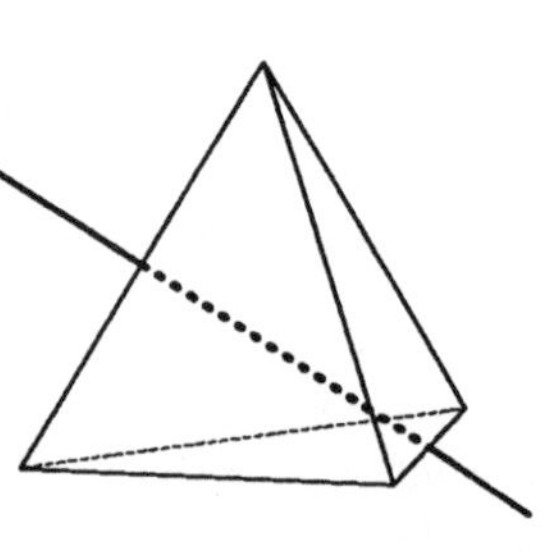

   a. Hold the straw vertically and look directly down on the model as you turn it. What kind (?-fold) of rotation symmetry does this axis demonstrate on the tetrahedron? *2-fold.*

      Cube? *2-fold.*

      Octahedron? *2-fold.*

   b. How many such axes of symmetry does the cube have?

      Six.

   c. How many such axes of symmetry does the octahedron have?

      Six.

   d. How many such axes of symmetry does the tetrahedron have?

      Three.

8. Summarize your findings about **rotation symmetry** by entering your answers to Exercises 2-7 in the following table.

Name	# of 4-fold Axes	# of 3-fold Axes	# of 2-fold Axes
Cube	3	4	6
Octahedron	3	4	6
Tetrahedron	0	4	3

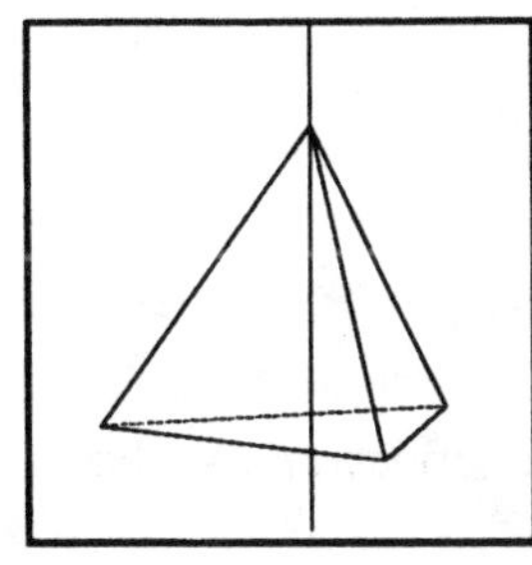

Activity 2 (continued)

Reflection Symmetry

A three-dimensional figure has **reflection symmetry** about a plane if the plane cuts the figure into two mirror-image shapes.

Each Platonic solid has several planes of reflection. In the exercises below, you will investigate the reflection planes of the cube and the regular octahedron. Build the transparent cube from an acetate net (net page 2), taping the edges with transparent tape. Use the regular octahedron from Activity 1, and colored rubber bands.

9. Demonstrate one plane of reflection of the cube by wrapping a rubber band *across* four parallel edges, as shown. The rubber band must bisect each face it crosses into two mirror-image halves.

The rubber band shows the intersection of the cube with one of its reflection planes.

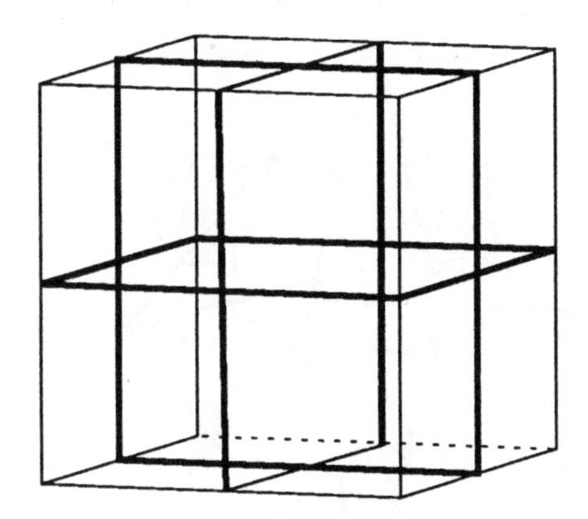

a. How many edges does the cube have?

12

b. How many of these reflection planes does the cube have?

Three.

c. Put additional rubber bands on your cube to show all these reflection planes, and sketch them on the picture.

10. The cube has another set of reflection planes. Demonstrate one such plane by wrapping a colored rubber band along two opposite edges and across two faces, as shown.

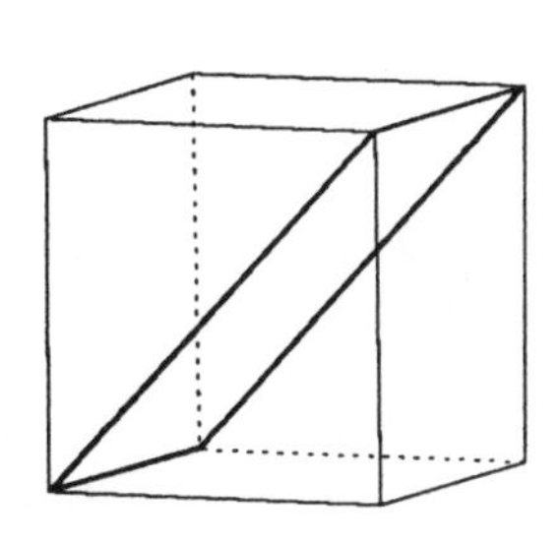

How many of these planes of reflection symmetry are there?

Six.

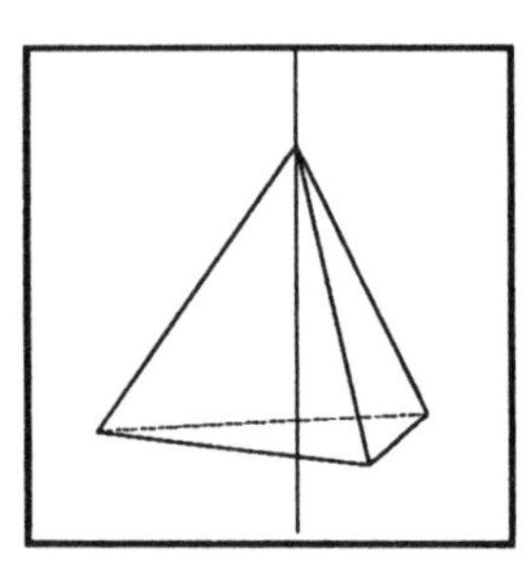

Activity 2 (continued)

11. Take all the rubber bands off, open the cube and place the octahedron inside so that each vertex of the octahedron touches the center of a face of the cube. Each plane of reflection for the cube will also be a plane of reflection for the octahedron.

    a. Place one rubber band on the cube as in Exercise 9. Sketch this reflection plane on the octahedron below.

    b. Place another rubber band on the cube as in Exercise 10. Sketch this reflection plane on the octahedron.

One kind of reflection plane of the octahedron cuts the shape into two pyramids. The pyramids of ancient Egypt were not exactly half-octahedra, however, as the slope of their walls was slightly less steep than the walls of a regular octahedron.

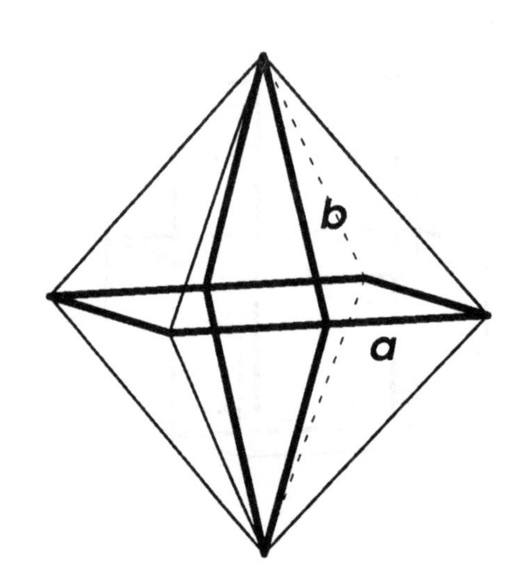

Teacher Comments for Activity 2

In this exercise, students will explore the rotation and reflection symmetries of three-dimensional figures. They will determine the rotation symmetries of the cube, the regular tetrahedron, and the regular octahedron, and the reflection symmetries of the cube and the octahedron.

Background Knowledge: Symmetry of two-dimensional figures (See Activity P-2).

Presenting the Activity: Compare rotation and reflection symmetries in three dimensions to the corresponding symmetries in two dimensions. The **center** of rotation (a **point**) in two dimensions becomes an **axis** of rotation (a **line**) in three dimensions; the axis of reflection in two-dimensions becomes a **plane** of reflection in three-dimensions.

Materials: Finished models of the Platonic solids from Activity 1; thin straws or pipe cleaners to serve as axes of rotation; transparent cube; rubber bands (size 14 or 16).

Building the transparent cube: An acetate cube net is included with the manipulative kit. If you use the blackline master net (net page 2), transfer it to a piece of 10 mil. acetate. Assemble the cube by taping the edges with transparent tape. (Acetate is available in many art and hobby supply stores and catalogs and in any plastics store or catalog.)

Comments on Activity Questions

1. Students may wonder what is so special about regular polyhedra; it's helpful to have some nonregular polyhedra (such as a rhombic dodecahedron) to compare to the Platonic solids.

2. In order to see that the faces and corners of the octahedron have actually changed position after a quarter turn, you may want students to put numbers on the faces of the octahedron. Ask how the numbers change position; also ask how edges and vertices change position.

4. The cube has four main diagonals; each is a 3-fold rotation axis.

5b. The octahedron has four pairs of opposite faces. Sit the octahedron flat on one of its faces to see that opposite faces are parallel.

6. Note that the 3-fold axes of the three different Platonic solids are of three distinct kinds: in the cube, a 3-fold axis joins a pair of vertices; in the octahedron, it joins the centers of a pair of faces; in the tetrahedron, it joins a vertex to the center of a face. All of the rotation axes go through the center of the solid (since a rotation of a polyhedron must leave the center fixed).

7. Emphasize that the axis must pass through the center of the polyhedron. This helps students line up the opposite edges.

7a. The edges pierced by the rotation axis must remain fixed by the rotation, so a 2-fold (180°) rotation is required.

7b. The number of 2-fold rotation axes is equal to half the number of edges of the polyhedron.

8. Note that the cube and the octahedron have the same number of each kind of rotation axes. To understand why, place the octahedron in the plastic cube so that each vertex of the octahedron touches the center of a face of the cube. Now you can see that any axis that passes through the centers of opposite faces of the cube also passes through opposite vertices of the octahedron. Note that any axis that passes through opposite vertices of the cube also passes through the centers of opposite faces of the octahedron.

Teacher Comments for Activity 2 (continued)

9b. Each reflection plane cuts four edges, so there are three such planes.

9c. Note that the three reflection planes are mutually perpendicular. Ask students to describe the number and shape of the pieces into which the cube is divided by one reflection plane; by two reflection planes; by three reflection planes of this type.

10. Each of these planes cuts through a pair of parallel edges; there are six such pairs. Again ask students to describe the number and shape of the pieces into which the cube is divided by reflection planes of this type. For more than one reflection plane the answers are not easy to imagine or to see. You might use a cube cut from firm clay, slicing it with a wire to outline the reflection planes.

When all reflection planes cut the cube, it is divided into 48 irregular tetrahedra, 24 of them congruent; the other 24 are their mirror images. The animated film, *Symmetries of the Cube*, shows this dissection.

11a. The three mutually perpendicular reflection planes of the regular octahedron show that its internal structure is that of the Cartesian coordinate system. Students can make an elegant model of these intersecting planes using the technique of unit origami. See Franco, *Using Tomoko Fusè's Unit Origami in the Classroom*, page 2.

Discussion/Extension: You cannot check for reflection symmetry in three dimensions by folding or turning one half over to see if it matches the other half, unless the object has rotation symmetry as well. For example, although a mirror image of a left shoe is a right shoe, you cannot turn one into the other by a physical motion.

Student Project 2 examines the rotation symmetry of the remaining two Platonic solids, the dodecahedron and the icosahedron. Student Project 3 uses mirrors to explore the reflection symmetry of all the Platonic solids. The film *Dihedral Kaleidoscopes* gives a colorful presentation of how symmetric polyhedra can be generated by mirrors from small "seeds."

Activity 3: Other Symmetric Polyhedra

A polyhedron with the following properties is **regular**:

a. All faces are regular polygons.

b. All faces are congruent to each other.

c. The same number of faces meet at each vertex in the same way.

The most familiar prism shape is a triangular prism. A glass triangular prism separates light into a rainbow spectrum of colors.

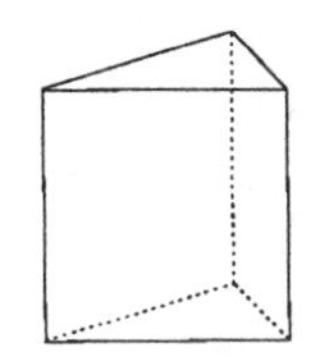

If a polyhedron lacks any of these properties, then it is not regular. In this activity you will build several polyhedra that are not regular.

1. Use one square panel and four triangle panels to construct a square pyramid, as shown in the picture.

   This polyhedron is not regular. Which properties above does it lack?

   b and c.

An antiprism has two congruent regular polygonal bases joined by equilateral triangles.

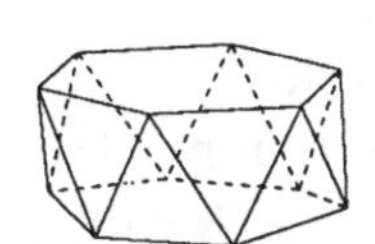

Drums are often laced like antiprisms. The top polygon is rotated with respect to the bottom polygon, giving the antiprism a twist.

2. A **regular right prism** is formed by two congruent regular polygonal bases which lie in parallel planes and have corresponding edges joined by squares that are perpendicular to the bases.

   Use two triangle panels as bases and connect them with three square panels to construct a regular right triangular prism.

   Use two pentagon panels as bases and connect them with five square panels to construct a right pentagonal prism.

   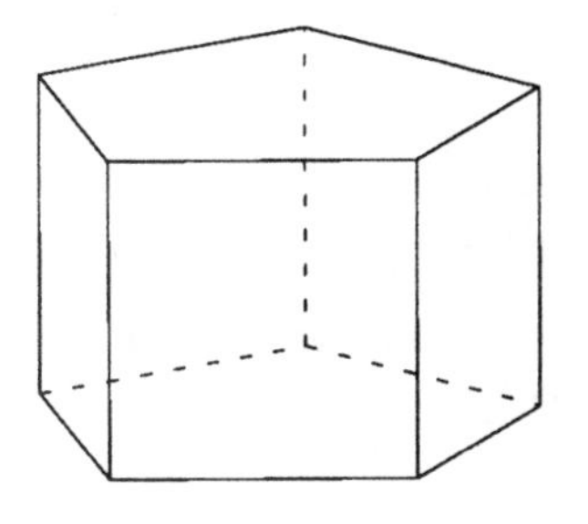

   a. The two prisms you have made are not regular polyhedra. Which property for regular polyhedra do they lack?

   b.

   b. One of the Platonic solids is also a right prism, but it is not usually called a prism. Which one is it?

   The cube (or regular hexahedron).

One of the Platonic solids is also an antiprism. Which one?

Activity 3 (continued)

If a polyhedron has properties a and c above but lacks property b, then it is called **semiregular**. All prisms and antiprisms are semiregular polyhedra. There are 13 other semiregular solids; these are known as the **Archimedean solids**. Like the Platonic solids, all vertices of an Archimedean solid are exactly the same, and the faces are regular polygons, but there are two or more different types of faces.

The Archimedean solids are named after Archimedes, a Greek who lived in the 3rd century B.C. He is the mathematician who, according to legend, jumped naked from the bathtub and ran through the streets shouting "Eureka!" when he discovered that a body placed in water will displace an amount of water equal to the weight of the body. Another legend says that he was killed by Roman soldiers when he refused to accompany them until he had finished the geometry problem on which he was working.

3. One Archimedean solid is the **cuboctahedron**. Build one by connecting two squares and two triangles, alternately, at each vertex. Continue until you have a closed shape.

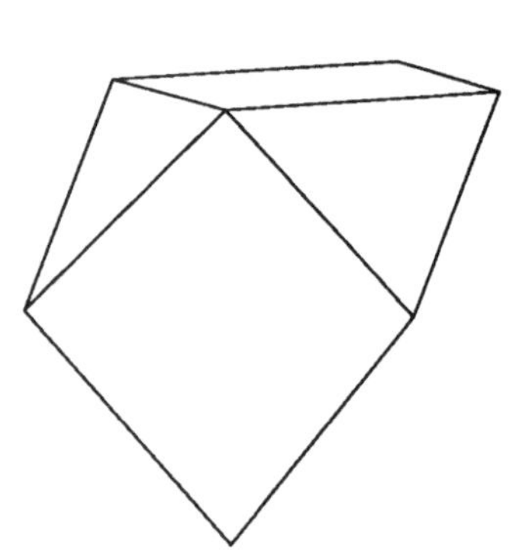

a. How many squares did you use altogether?

Six.

b. How many triangles did you use altogether?

Eight.

c. Explain why the shape has the name cuboctahedron.

Answers will vary.
See Teacher Comments.

4. For another example of a polyhedron that is not regular, assemble 12 rhombus panels as shown in the diagram. Note that the angles that meet at each vertex are congruent to each other. The shape you have made is called a **rhombic dodecahedron**.

The cuboctahedron is an Archimedean solid. Another one is the truncated icosahedron, which is pictured at the top of the page.

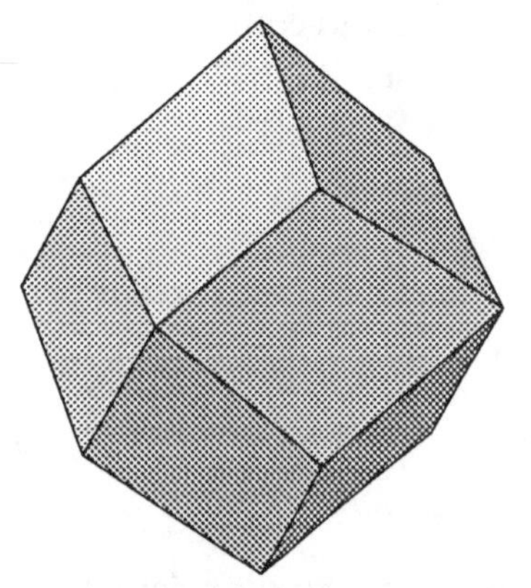

a. Which properties of a regular polyhedron does this shape lack?

a and c.

If you pack twelve congruent spheres around a single congruent sphere, as if you were making a pile of snowballs, the centers of the outside spheres will form a cuboctahedron.

b. Is it a semiregular polyhedron? Explain.

No. The vertices are not the same and the faces are not regular polygons.

c. Explain why the shape has the name *rhombic dodecahedron*.

It has 12 faces, and each face is a rhombus.

Teacher Comments for Activity 3

Using polygon panels, students will construct various polyhedra that are not regular, yet satisfy some of the properties of regular polyhedra. These include pyramids, prisms, Archimedean solids, and duals of Archimedean solids.

Background Knowledge: Congruence; regular polygons; cube; octahedron; rhombus; and dodecahedron.

Presenting the Activity: Review the three properties of a regular polyhedron. Remind students that all three properties are necessary in order to have a regular polyhedron. This is a good activity to do in groups, as considerable building is required.

Materials: 14 triangle panels, 15 square panels, 2 pentagon panels, 12 rhombus panels; rubber bands size 10.

Comments on Activity Questions

1. Square pyramids are familiar shapes. (The 1990 addition to the Louvre museum in Paris is a square pyramid.) There are an infinite number of different pyramids; in each, the base is a regular polygon, and the remaining faces are triangles. W. Blackwell, *Geometry in Architecture,* has a chapter on pyramids and cones and another on prisms and antiprisms.

2. The use of right prisms to produce the spectrum of colors from white light is very important in optics.

   There are an infinite number of regular right prisms, named according to the shape of their bases. It is relatively easy to find the surface area and volume of a right prism. Basically, they are just thick polygons. To calculate the volume, simply multiply the area of one base by the height of the prism; to calculate surface area, multiply the perimeter of one base by the height of the prism.

3. See the Glossary for drawings of the 13 Archimedean solids. The snub cube and the snub dodecahedron appear twisted. In fact, each can take two forms—with a right- or left-handed twist. These mirror images are called enantiomorphs. Some references claim there are 14 Archimedean solids, and include a "wrongly assembled" rhombicuboctahedron having one triangle and three squares at each vertex. This polyhedron is obtained by twisting a "cap" of the rhombicuboctahedron 1/8 of a turn, and has less symmetry than the rhombicuboctahedron. See P. O'Daffer and S. Clemens, *Geometry: An Investigative Approach,* p. 141, and also H.M. Cundy and A. Rollett, *Mathematical Models,* section 3.7.14, on **isomerism**.

3c. There are several possible explanations. The cuboctahedron has six square faces like a cube and eight triangular faces like an octahedron. The cuboctahedron results from a truncation of either a cube or an octahedron (see Activity 7). The cuboctahedron is the core of the interpenetrating duals, the cube and the octahedron (see Student Project 12).

4. In forming the corners of the rhombic dodecahedron, be sure to put congruent angles together.

4a. This polyhedron has only one face shape—the rhombus, which is not a regular polygon. All vertices are not the same—three faces meet at some vertices, while four faces meet at others.

Teacher Comments for Activity 3 (continued)

Discussion/Extension: All of the nonregular polyhedra in this activity have some symmetries, though they are not symmetric in as many ways as the Platonic solids. Use the same techniques as in Activity 2 to investigate rotation and reflection symmetry. The symmetries of the cuboctahedron and the rhombic dodecahedron are the same because these figures are duals (see Activity 6).

See the Glossary for pictures of the Archimedean solids. See Student Project 4: Truncated Octahedron, and Student Project 9: Truncated Icosahedron, for two examples of Archimedean solids. Student Project 5 has more activities on the rhombic dodecahedron.

There are some other interesting nonregular polyhedra. The stella octangula is an example of a nonconvex polyhedron that satisfies properties a and b, but not c. (See Visual Geometry Project, *The Stella Octangula.*) You can make a very simple nonregular tetrahedron that satisfies properties b and c, but not a, by folding up a net consisting of a strip of four congruent acute (but not equilateral) triangles.

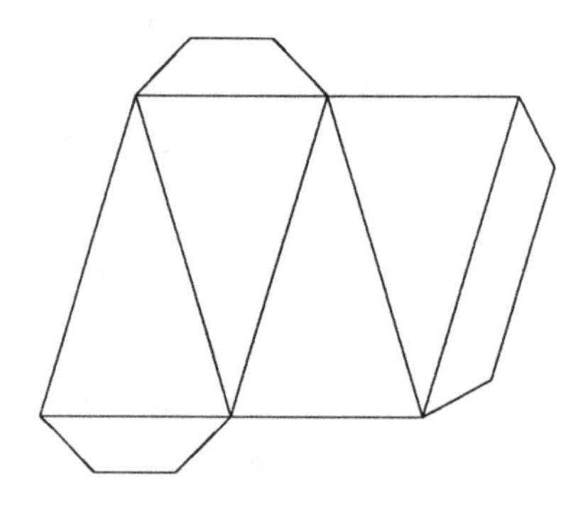

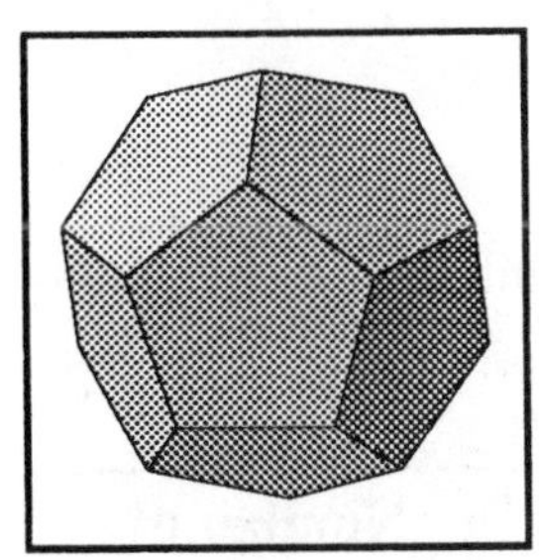

Activity 4: Counting Faces, Edges, and Vertices of Polyhedra

1. Look at the regular dodecahedron. How many faces does it have?

   Twelve.

   In order to find the number of *edges* on the regular dodecahedron, you could just count them. But the regularity of this polyhedron allows us to use a shortcut.

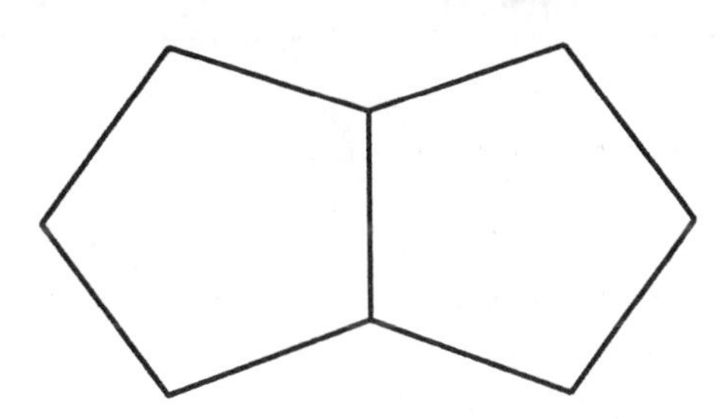

   Note that there are 5 edges on each face, and that exactly 2 faces share an edge. Therefore, there are (12 x 5)/2 = 30 edges on the regular dodecahedron.

The angle between two faces of a polyhedron is called a ***dihedral angle****. All dihedral angles of a regular polyhedron have the same measure.*

When you count each edge on each face of a polyhedron, you are counting each edge of the polyhedron twice.

When you count each vertex on each face of a polyhedron, you are counting each vertex of the polyhedron more than once, depending on the number of faces which meet at each vertex.

2. You can use a similar shortcut to find the number of *vertices* on the regular dodecahedron.

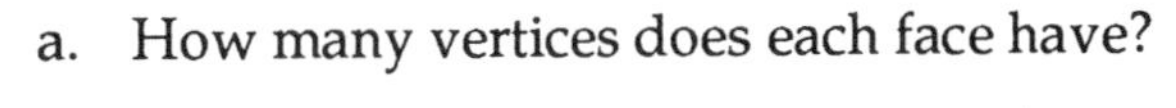

   a. How many vertices does each face have?

      Five.

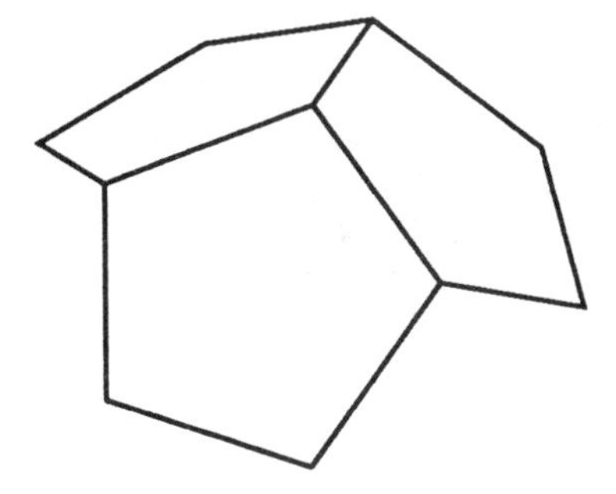

   b. How many faces meet at each vertex of the dodecahedron?

      Three.

   c. How many vertices are there on the regular dodecahedron?

      (12 x 5)/3 = 20

3. Enter the information from Exercises 1 and 2 in the chart below. Fill in the chart in a similar manner for the rest of the regular polyhedra that you have built.

Name	Number of Faces (f)	Number of Vertices (v)	Number of Edges (e)
Tetrahedron	4	4	6
Hexahedron (cube)	6	8	12
Octahedron	8	6	12
Dodecahedron	12	20	30
Icosahedron	20	12	30

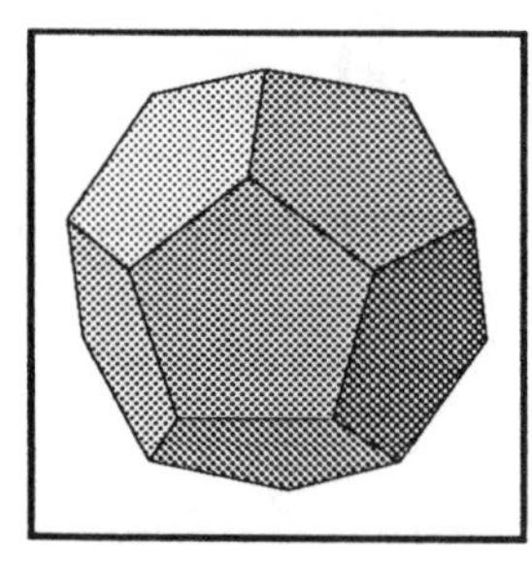

Activity 4 (continued)

4. Fill in the chart below for the polyhedra listed. When possible, find shortcuts to count edges and vertices.

Name	Number of Faces (f)	Number of Vertices (v)	Number of Edges (e)
Square Pyramid	*5*	*5*	*8*
Right Triangular Prism	*5*	*6*	*9*
Right Pentagonal Prism	*7*	*10*	*15*
Cuboctahedron	*14*	*12*	*24*
Rhombic Dodecahedron	*12*	*14*	*24*

Numbers alone do not determine the shapes of polyhedra. For example, a pentagonal prism has exactly the same number of faces, vertices, and edges as the shape which is obtained by cutting off one corner of a cube.

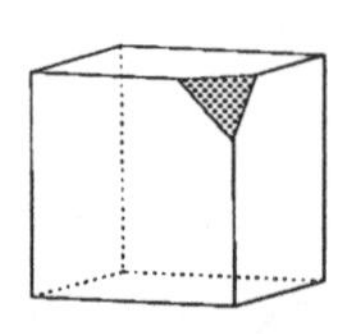

However, this shape differs considerably from the pentagonal prism in the shape of the faces and the way in which the faces are connected.

5. There are several patterns in the charts in Exercises 3 and 4. Describe a pattern that you see.

Answers will vary. See Teacher Comments.

Teacher Comments for Activity 4

Students will learn to count faces, edges and vertices of polyhedra in a way that is efficient and exploits the regularity of the solids. As results are tabulated, patterns will emerge. Among those patterns will be the discovery of duality and Euler's Formula.

Background Knowledge: Definitions of vertices, edges, faces, and polyhedra.

Presenting the Activity: It might be best for students to work in groups since the activity calls for examining ten different polyhedra. Point out to the students that while it may be easy to directly count the faces, edges and vertices of some of the Platonic solids, it is not so easy for more complicated polyhedra. It's helpful to learn the process described in the activity on the simpler solids first.

Materials: Models of the five regular polyhedra; square pyramid; triangular prism; pentagonal prism; cuboctahedron; rhombic dodecahedron.

Comments on Activity Questions

1. In the calculation (12 x 5) = (no. of faces x no. of edges per face), each edge of the polyhedron is counted twice, since each edge is shared by two faces. Therefore, we must divide by 2.
2. In the calculation (12 x 5) = (no. of faces x no. of vertices per face), each vertex of the polyhedron is counted 3 times, since each vertex is shared by 3 faces. Therefore, we must divide by 3: there are (12 x 5)/3 = 20 vertices.
3. Have students explain the process of counting edges and vertices as in Exercises 1 and 2.
4. Have students describe their shortcuts. The process in Exercises 1 and 2 depends on the complete regularity of the polyhedra, and must be modified for this exercise. For example, the number of edges of the cuboctahedron is [(6 x 4) + (8 x 3)]/2 = 24. You must calculate separately for square and triangular faces.

   Shortcuts for counting vertices and edges on the prisms are easy to find. Note that for any prism with an n-gon base, $f = n + 2$, $v = 2n$, and $e = 3n$.
5. The sum of the faces and vertices is always 2 less than the number of edges. This is Euler's formula (see Activity 5).

   The number of edges for some pairs of polyhedra are the same, and for these pairs, the numbers of faces and vertices are interchanged. This hints at duality, which is further explored in Activity 6.

Discussion/Extension: Count the edges and vertices of the other Archimedean solids using the shortcuts of this activity (see Glossary for pictures of the Archimedean solids). If you have a model of the stella octangula (see Visual Geometry Project, *The Stella Octangula*), try counting its edges and vertices.

The methods of counting in this activity can also be used to estimate the number of vertices and edges on a polyhedron when precise information is not known (see Activity 5). Map-coloring consequences using this kind of counting can be found in S. Stein, *Mathematics, the Man-Made Universe*, Chapter 13. Examples of pentagonal dodecahedra in nature are illustrated in J. Galloway, "Nature's Second Favourite Structure," *New Scientist*, 31 March 1988.

Activity 5: Euler's Formula and Other Relationships

Leonard Euler (pronounced "Oi•ler"), an 18th century Swiss mathematician, was one of the most productive mathematicians of all time. Euler loved children and often worked with a baby on his lap and some of his other children playing around him (he had 13, though 8 of them died while still young).

1. For each regular polyhedron, add the number of faces (f) and the number of vertices (v), and compare this sum to the number of edges (e). Complete the following equation to describe the relationship:

$$f + v = e + 2$$

This relationship is known as **Euler's Formula.**

2. Write down the names of some other polyhedra which fit Euler's Formula. (Check the ones in the second chart in Activity 4. Perhaps you can find others, as well.)

Square pyramid, right triangular prism, right pentagonal prism, cuboctahedron, and rhombic dodecahedron. Students may also find others.

3. Euler's Formula applies to any polyhedron which, if made of rubber, could be blown up into a sphere. Euler's Formula does *not* apply to polyhedra which have tunnels or holes in them. Demonstrate this by finding the numbers of faces, vertices, and edges for the polyhedron drawn below. (Its faces are all trapezoids or rectangles.)

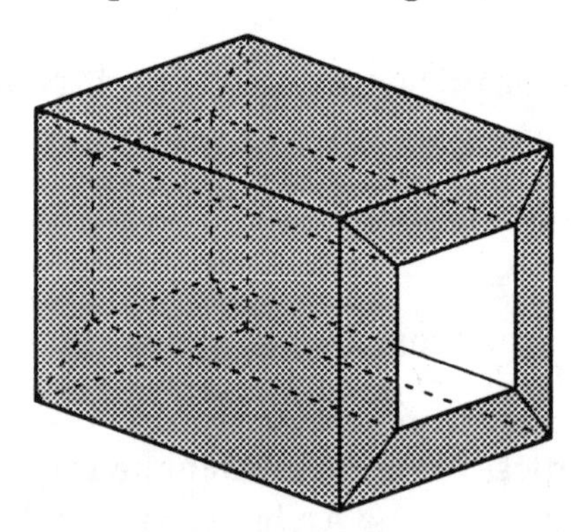

$f =$ *16*

$v =$ *16*

$e =$ *32*

A hat, called a mazzocchio, had the shape of a polyhedron doughnut and was worn by men in the time of the Renaissance.

4. The picture on the right shows the microscopic sea creature called *aulonia hexagona.* It is often described as having hexagonal cells with three hexagons meeting at each vertex. Look carefully at the picture, and you'll see that some of the cells are not hexagons. What other shapes do you see?

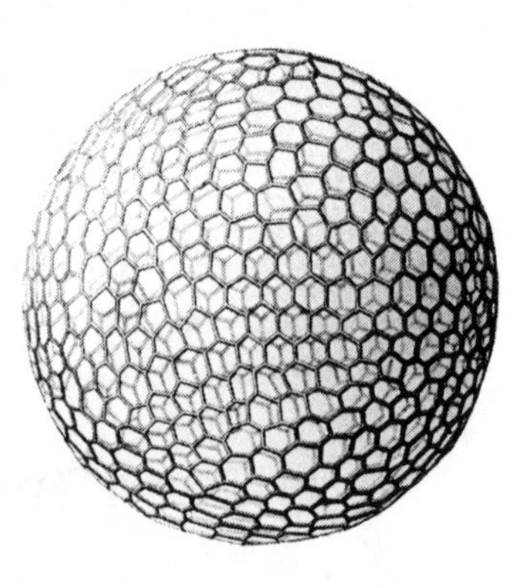

Activity 5 (continued)

5. Euler's Formula can be used to prove that there is no convex polyhedron with just hexagonal faces, three meeting at each vertex.

   For the proof, we suppose there is such a polyhedron and then show that it's impossible.

Remember that when you multiply the number of edges on each face by the number of faces, you have counted each edge of the polyhedron twice.

   a. Suppose there is a convex polyhedron with only hexagonal faces, three meeting at each vertex. Let f be the number of faces on this polyhedron. Find an expression for the number of edges (e) in terms of f.

      $e = 6f/2 = 3f$

   b. Find an expression for the number of vertices (v) in terms of the number of faces (f).

      $v = 6f/3 = 2f$

   c. If the polyhedron is convex, Euler's Formula must be true for the faces, vertices, and edges of the polyhedron. Use the expressions obtained in a and b to substitute for v and e in Euler's Formula. Is the equation satisfied?

      $f + v = e + 2$
      $f + 2f = 3f + 2$
      This give the contradiction 0 = 2. Therefore, the equation is not satisfied.

   When a polyhedron does not satisfy Euler's Formula, it cannot be convex.

6. Even when you cannot directly count faces, edges, and vertices of a polyhedron, the properties of the polyhedron determine certain relationships between these numbers. Here's one such relationship which is true for *any* polyhedron with f faces, v vertices, and e edges: $2e \geq 3f$.

   Here's how to obtain this inequality:

   a. Each face of the polyhedron must have three or more edges. Why?

      A triangle is the polygon with the fewest edges.
      Every polygon must have three or more edges.

Activity 5 (continued)

b. Add up all the edges of all the faces; the sum of all these edges is greater than or equal to $3f$. But this sum also equals twice the number of edges on the polyhedron. Why?

Because each edge is joined to exactly two faces.

Therefore, $2e \geq 3f$.

c. Check that this inequality holds for the polyhedra that you have built. For which of the Platonic solids is $2e$ exactly equal to $3f$?

$2e = 3f$ for the tetrahedron, the octahedron, and the icosahedron.

The inequalities $2e \geq 3f$ and $2e \geq 3v$, together with Euler's Formula, can be used to show that every convex polyhedron must have at least 4 faces, 4 vertices, and 6 edges. Thus, the tetrahedron is the "minimum" polyhedron.

Similarly, no convex polyhedron with 7 edges can exist. The three relationships here can be used to prove this.

7. A similar inequality gives a relationship between the number of edges e and the number of vertices v of any polyhedron.

a. In any polyhedron, at least three edges meet at each vertex. Why?

At least three polygons must meet at a vertex in order to form a three-dimensional shape.

b. Add up all the edges that meet at each of the vertices; the sum of all these edges is greater than or equal to $3v$. But this sum also equals twice the number of edges on the polyhedron. Why?

Because each edge is counted by two different vertices.

Therefore, $2e \geq 3v$.

c. For which of the Platonic solids is $2e$ exactly equal to $3v$?

$2e = 3v$ for the tetrahedron, the cube, and the dodecahedron.

Teacher Comments for Activity 5

Euler's formula, $f + v = e + 2$, holds for all convex polyhedra. We do not prove it here, but students will verify Euler's formula for the Platonic solids and several nonregular polyhedra. We also give an example of a polyhedron which does not adhere to Euler's formula.

Other relationships apply to the numbers of faces, vertices, and edges of polyhedra. In this activity, we derive the inequalities $2e \geq 3f$ and $2e \geq 3v$, which hold for any polyhedron.

Background Knowledge: Activity 4 should be completed before or combined with this activity.

Presenting the Activity: Review the counting process used in Activity 4 for counting edges and vertices of regular polyhedra. Have students describe the counting process in words.

Materials: Students will need their filled-in charts from Activity 4 and/or models of regular and nonregular polyhedra.

Comments on Activity Questions

1. A special issue of *Mathematics Magazine* (Nov. 1983) is devoted to the life and work of Euler.

   Have students describe some triples of numbers (v, e, f) that *cannot* be the number of vertices, edges, and faces of a convex polyhedron because they do not satisfy Euler's formula (any set of three odd positive numbers, for example).

3. Student Project 10 gives instructions for building a polyhedron with a hole, known as a toroid. Although the form of Euler's formula given here does not apply to toroids, there is a special case of Euler's formula for shapes with holes: if the shape has h holes, then the general formula is $f + v = e + 2 - 2h$.

5. This technique of "supposing" is the indirect method of proof.

5a. Count as in Activity 4, Exercise 1.

5b. Each edge joins exactly two vertices, so each edge is counted by the two vertices it joins.

6c. Equality holds when all the faces of the polyhedron are triangles, since a triangle is the polygon with the fewest number of sides.

7c. Equality holds when exactly three faces meet at each vertex, since three is the minimum number of faces that can meet at a vertex.

Discussion/Extension: There are several ways to prove Euler's theorem. The usual way is to view the three-dimensional polyhedra as two-dimensional networks of regions, edges, and vertices. See, for example, A. Beck, M. Bleicher, and D. Crowe, *Excursions into Mathematics*, pp. 8-11, and S. Stein, *Mathematics, the Man-Made Universe*, pp. 350-351. For an enjoyable proof using dikes and flooded fields, see H. Rademacher and O. Toeplitz, *The Enjoyment of Mathematics*, p. 75.

Many conclusions can be drawn by making algebraic deductions using counting arguments, as was done in this exercise. Examples of these, including the proof that no convex polyhedron with exactly 7 edges can exist, are provided in Student Project 7, Applications of Euler's Formula, Student Project 8, Deltahedra, and Student Project 9, The Truncated Icosahedron. More examples appear in Beck et al., Chapter 1, H. Rademacher and O. Toeplitz, pp. 73-78, D. Schattschneider, "Counting it Twice," and J. Pedersen and P. Hilton, *Build Your Own Polyhedra*, pp. 159-166.

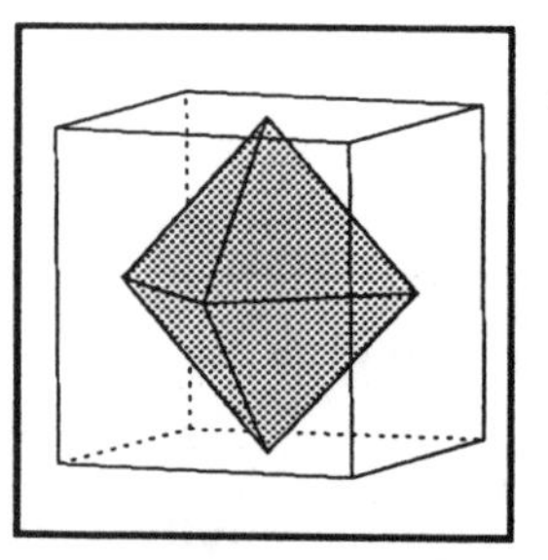

Activity 6: Duality of Polyhedra

1. a. How many faces does the cube have? *Six.*

   How many vertices does the octahedron have? *Six.*

   b. How many faces does the octahedron have? *Eight.*

   How many vertices does the cube have? *Eight.*

*A polyhedron is **inscribed** in a solid if it is inside the solid and each vertex of the polyhedron touches the surface of the solid.*

One polyhedron is the **dual** of another if its faces correspond in a special way to the vertices of the other and its vertices correspond similarly to the faces of the other. For example, the cube and the octahedron are duals of each other.

In the next exercise, you will demonstrate the dual relationship of the cube and the octahedron.

2. Open the transparent cube and place the octahedron inside so that each vertex of the octahedron touches the center of a face of the cube.

   a. Describe the position of the edges of the octahedron in relation to the cube.

   The edges of the octahedron connect the centers of the faces of the cube.

   b. Tilt the cube so that you are looking directly down at one corner. Describe the position of the vertex at that corner in relation to the closest face of the octahedron inside.

   The vertex of the cube is directly above the center of the triangular face of the octahedron.

One way to find the dual of a Platonic solid or an Archimedean solid is to connect the centers of adjacent faces of the solid. This produces a model of its dual inscribed in the polyhedron, just like your octahedron inside the transparent cube.

3. Let's try this on the octahedron.

The dual of the dual of a polyhedron is the original polyhedron; that is, if one polyhedron is the dual of a second polyhedron, then the second is the dual of the first. Each Platonic solid is the dual of another Platonic solid.

   a. How many faces meet at each vertex of the octahedron? *Four.*

   b. One face of the dual of the octahedron is made by connecting the centers of adjacent faces that surround a vertex of the octahedron. Use segments to connect the centers which have been marked in the picture. *Connect the centers of adjacent faces only.* What shape have you made?

   A square.

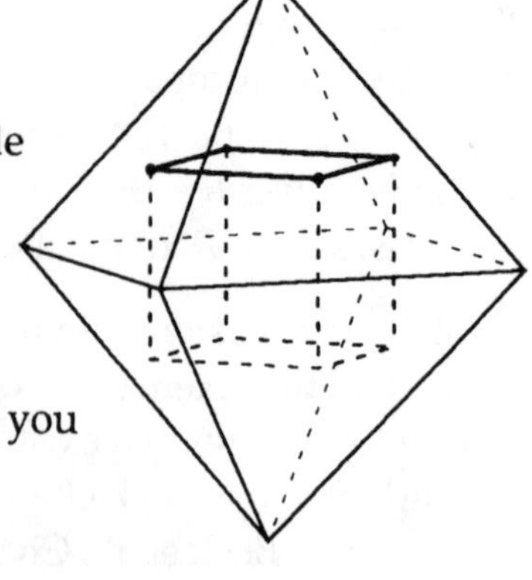

   c. On the picture of the octahedron, draw the rest of its inscribed dual.

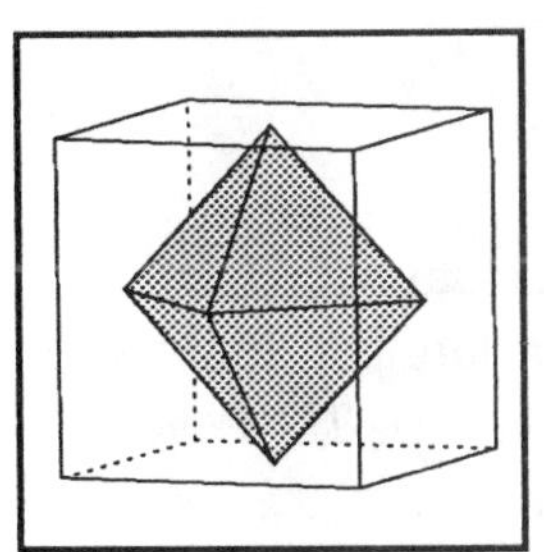

Activity 6 (continued)

4. The picture below shows one corner of a regular dodecahedron.

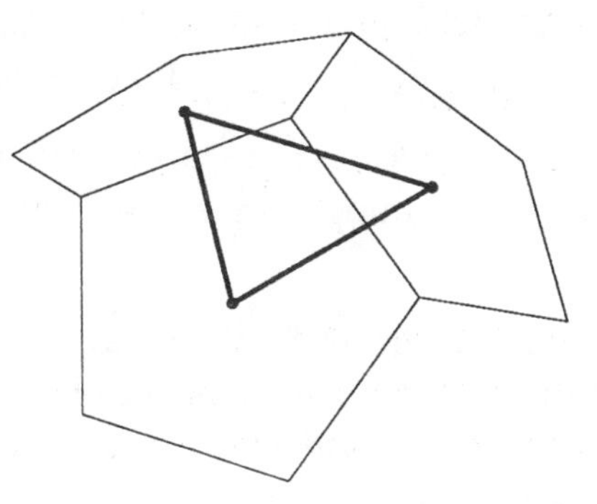

   a. On the picture, sketch one face of the dual figure. What polygon is this?

      An equilateral triangle.

   b. How many vertices does the dodecahedron have?

      20

   c. How many faces will the dual have?

      20

   d. What polyhedron is the dual of a regular dodecahedron?

      A regular icosahedron.

5. What polyhedron is the dual of a regular icosahedron? Explain.

   A regular dodecahedron. See Teacher Comments.

6. What can you say about the number of edges of a polyhedron and its dual?

   A polyhedron and its dual have the same number of edges.

Archimedean solids have two or more kinds of regular polygon faces, but have identical vertices. Duals of Archimedean solids have congruent faces (why?) but are not regular polyhedra (why not?).

7. Draw the inscribed dual of the tetrahedron on the picture. What polyhedron is the dual of the regular tetrahedron?

   Another regular tetrahedron.

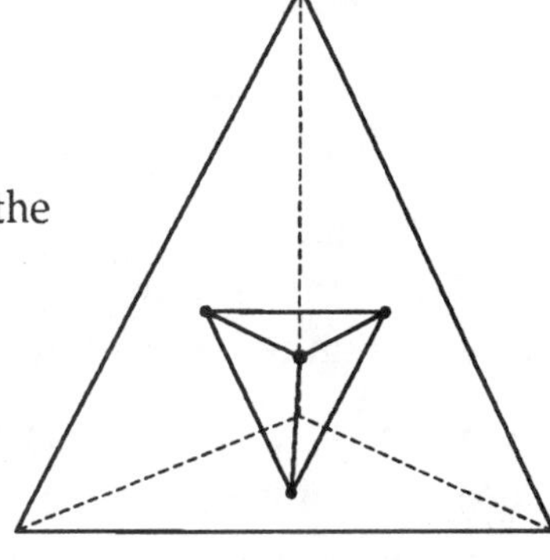

8. a. Examine the second chart in Activity 4 to find a pair of non-regular polyhedra that might be duals of each other. How can you test that the two polyhedra are duals?

      The cuboctahedron and the rhombic dodecahedron. Connect the centers of the faces of one polyhedron to see if you get the other one.

   b. Find a polyhedron that is not regular and is dual to itself; that is, the shape of its dual is the same as the shape of the polyhedron.

      A square pyramid.

Teacher Comments for Activity 6

This activity is an examination of duality in three dimensions. This is related to the duals of two-dimensional tessellations, described in Activity P-3. Here, students will manipulate the dual pair of cube and octahedron and describe the duals of the other Platonic solids. The videotape *The Platonic Solids* gives an animated presentation of the concept of duality for all the Platonic solids.

Background Knowledge: The Platonic solids—number of faces, edges, and vertices of each one. For Exercise 8, students need the chart from Activity 4 containing the numbers of faces, edges, and vertices of nonregular polyhedra.

Presenting the Activity: Compare duality in three dimensions to the process of finding the dual of a tessellation, as in Activity P-3. Emphasize that the midpoints of adjacent faces are connected; this is how to find the dual of a polyhedron. The videotape *The Platonic Solids* gives a clear visual presentation of duality. You may want to show this before and/or after doing this activity.

Materials: Octahedron from Activity 1; transparent cube.

Comments on Activity Questions

2a. Note that the way in which the edges of the octahedron connect the midpoints of adjacent faces of the cube is analogous to how one draws the dual of a tessellation of regular polygons. This technique of obtaining the dual is graphically demonstrated in the Visual Geometry videotape *The Platonic Solids*. This technique doesn't always work on an arbitrary polyhedron because connecting midpoints of adjacent faces may not produce planar polygons. However, the regularity of the Platonic and Archimedean solids and the simplicity of some other polyhedra make the technique work for these shapes. Another way to visualize the dual relationship is to interpenetrate a polyhedron and its dual. This is developed in Student Project 12, and is also presented in the videotape *The Platonic Solids*.

2b. This demonstration shows that the vertices of the cube correspond in a natural way to the faces of the octahedron.

3c. Since the faces of the octahedron are distorted by the two-dimensional drawing, it is not easy to find the centers of the faces by eye. They can be found by drawing two medians on each face; the intersection of the medians is the center of the face. (Midpoints of edges are not distorted.)

4a. As in 3c, the center of each pentagon face can be found as the intersection of two medians in each face. (Join a vertex to the midpoint of the opposite side.)

5. Repeat the analysis of Exercise 4. One face of the dual figure will be a pentagon. The icosahedron has 12 vertices, so the dual will have 12 faces. Notice that taking the dual of the dual of the dodecahedron gives the dodecahedron again.

6. Since the number of vertices and faces are interchanged for a polyhedron and its dual, the number of edges of the two polyhedra must be equal, by Euler's formula. (When the numbers v and f are interchanged in the formula $v + f = e + 2$, the right-hand side does not change, so e stays the same.)

7. Remind students that the tessellation of squares is the two-dimensional dual of itself; the tetrahedron is a three-dimensional example of a self-dual figure.

Teacher Comments for Activity 6 (continued)

8a. The numbers of vertices and faces for the cuboctahedron and the rhombic dodecahedron suggest duality. Have students explain how this can be verified by beginning with one polyhedron and connecting centers of its adjacent faces to obtain the other polyhedron. This verification is necessary because numbers alone do not determine the shapes of faces or how they are connected. For example, the numbers of edges, faces, and vertices of a cube with one corner truncated are the same as for a pentagonal prism, but the duals of these polyhedra are entirely different.

8b. Any pyramid is self-dual; the regular tetrahedron is a special case.

Discussion/Extension: Have students explore the duals of other Archimedean solids and other simple polyhedra with regular faces (such as prisms) using the method of connecting centers of adjacent faces. Find some convex solids for which this method fails to produce a model of a solid with flat polygonal faces. Keep in mind that it is easy to find the centers of certain faces: regular polygons, arbitrary triangles (intersect medians), and parallelograms (intersect diagonals), for instance.

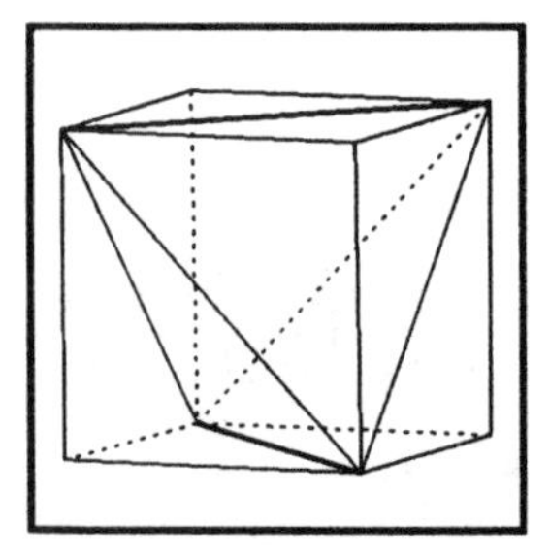

Activity 7: Inscribing Polyhedra in Each Other

A polyhedron is **inscribed** in a solid if it is inside the solid and each vertex of the polyhedron touches the surface of the solid.

Each of the Platonic and Archimedean solids can be inccribed in a sphere.

1. To inscribe a polyhedron in a solid, certain points on the surface of the solid are connected by line segments. For example, we can inscribe a cube in a dodecahedron by connecting a pair of non-adjacent vertices on each face of the dodecahedron.

   a. How many faces does the dodecahedron have?

   12

   b. How many edges does the cube have?

   12

   c. Each edge of the inscribed cube is a diagonal of one pentagon face of the dodecahedron. How many different diagonals does one face of the dodecahedron have?

   Five.

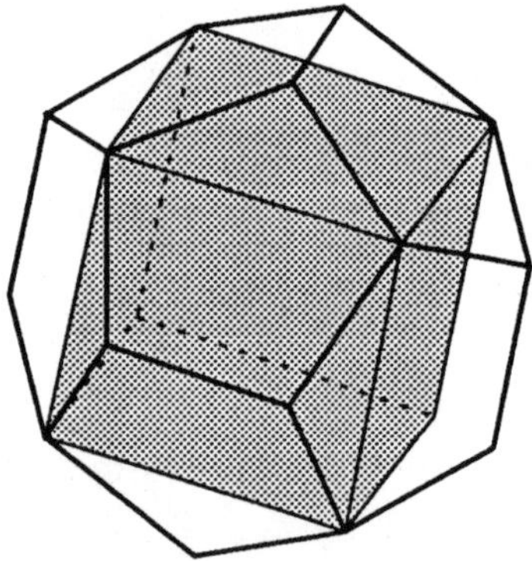

A compound of five interpenetrating cubes inscribed in a single dodecahedron is a beautifully symmetric star-like polyhedron.

   d. In how many different positions can a cube be inscribed in a dodecahedron?

   Five.

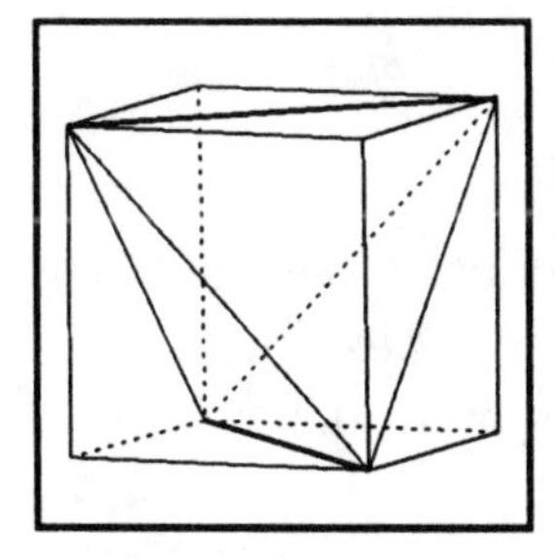

Activity 7 (continued)

2. In a similar fashion, a regular tetrahedron can be inscribed in a cube. You can see this by connecting certain non-adjacent vertices on the transparent cube. Use a non-permanent marker. First draw a diagonal on the top face. Now draw the *other* (non-parallel) diagonal on the bottom face. The endpoints of these two diagonals determine the four vertices of the tetrahedron. Connect the four vertices with line segments. Each edge of the tetrahedron will be a diagonal of a face of the cube.

   a. Where are the vertices of the tetrahedron in relation to the cube?

      They are at four vertices of the cube.

   b. In how many different positions can a tetrahedron be inscribed in a single cube in this way?

      Two.

Since a regular tetrahedron can be inscribed in a cube, and a cube can be inscribed in a regular dodecahedron, imagine how a regular tetrahedron could be inscribed in a regular dodecahedron. The vertices of the tetrahedron will touch the vertices of the dodecahedron.

3. Estimate the volume of the inscribed tetrahedron compared to the volume of the cube.

   Answers will vary.

4. (*optional*) Check your estimate by building a regular tetrahedron and four "cube corners" (net page 7).

   a. Put two cube corners together to form a pyramid with the same base and the same height as the regular tetrahedron. How does the volume of one cube corner compare to the volume of the regular tetrahedron?

      The volume of one cube corner is equal to half the volume of the tetrahedron since the two triangles have equal bases and equal heights.

   b. Fit all five models into the cube so that they are completely contained. Do you still believe your estimate in Exercise 3? Why or why not?

      The volume of the inscribed tetrahedron is 1/3 the volume of the cube.

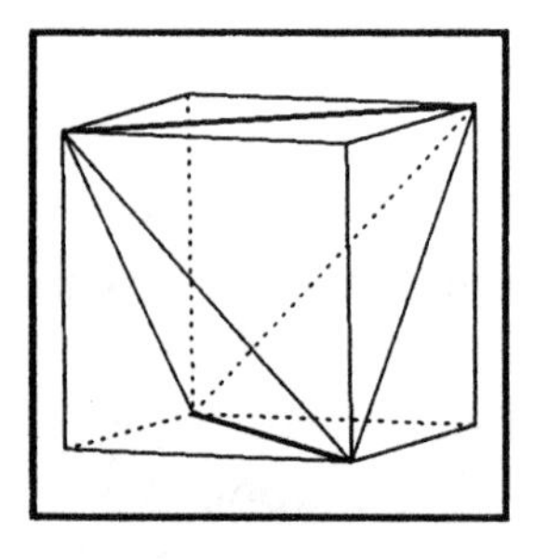

Activity 7 (continued)

5. Wipe off the marks you have made on your transparent cube. Now draw line segments connecting the midpoints of *adjacent* edges of the cube. The inscribed solid you have created, which is called a cuboctahedron, has both square and triangular faces.

The cuboctahedron can be described as a cube whose corners have been removed by slicing through the midpoints of adjacent edges. The cuboctahedron can also be thought of as an octahedron whose corners have been sliced off in the same manner.

   a. How many square faces does it have?

      Six.

   Where are the square faces of the cuboctahedron positioned in relation to the cube?

      They are centered on the six faces of the cube, but are turned 45° with respect to the cube faces.

   b. How many triangular faces does the cuboctahedron have?

      Eight.

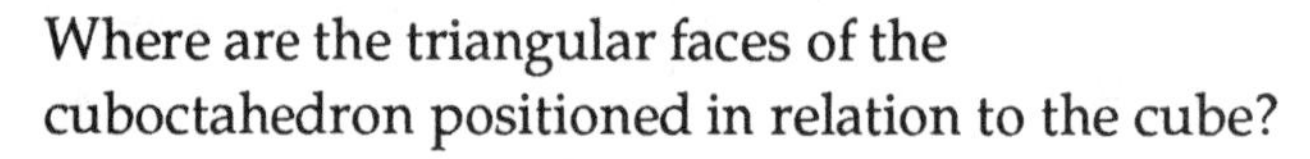

   Where are the triangular faces of the cuboctahedron positioned in relation to the cube?

      They are at the eight corners of the cube.

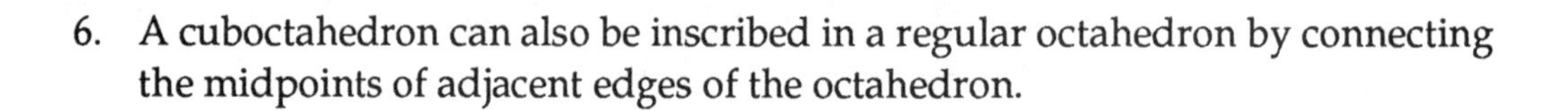

6. A cuboctahedron can also be inscribed in a regular octahedron by connecting the midpoints of adjacent edges of the octahedron.

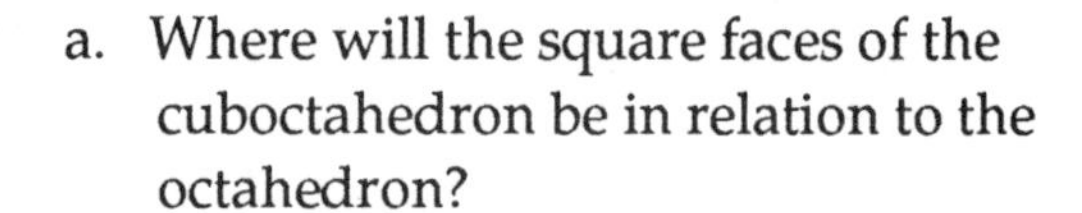

   a. Where will the square faces of the cuboctahedron be in relation to the octahedron?

      At the six corners of the octahedron.

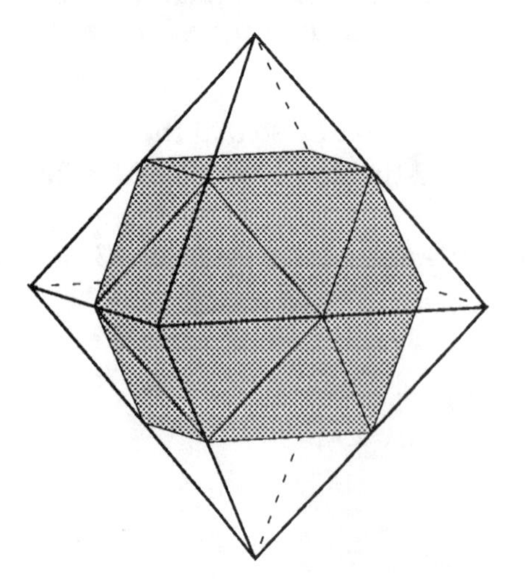

   b. Where will the triangular faces of the cuboctahedron be in relation to the octahedron?

      Centered on the eight faces of the octahedron.

Teacher Comments for Activity 7

This activity helps build spatial relations skills. Students determine how certain polyhedra can be inscribed in Platonic solids. These relationships are not immediately obvious. Students discover that a cube can be inscribed in a dodecahedron, and a tetrahedron can be inscribed in a cube. In each of these cases, the edges of the inscribed solid are diagonals of the faces of the outer polyhedron. This activity also discusses inscribing a cuboctahedron in a cube and in an octahedron.

Background Knowledge: Platonic solids, cuboctahedron.

Presenting the Activity: See A. Holden, *Shapes, Space, and Symmetry*, pp. 29-32, for pictures of various inscribed polyhedra. These may help students visualize the inscribed polyhedra, or they may be shown after students have done the activity.

Materials: Transparent cube; non-permanent (water-based) marker. Optional: tetrahedron net (net page 7); 4 cube corner nets (net page 7); rice, birdseed, or similar material.

Comments on Activity Questions

1c. For pictures of the compound of all five cubes interpenetrating each other as they would if inscribed in a single dodecahedron, see H. Cundy and A. Rollett, *Mathematical Models*, or A. Holden, *Shapes, Space, and Symmetry*. Many science museums have models of such compounds. Kits for building this and other complex compound polyhedra are available from *Symmetrics*.

2. Note that the tetrahedron has six edges, and the cube has six faces.

2b. The compound of two interpenetrating tetrahedra inscribed in a single cube is called a stella octangula. See Visual Geometry Project, *The Stella Octangula*.

You can combine exercises 1 and 2 by asking students to describe how to inscribe a regular tetrahedron in a dodecahedron so that its vertices coincide with vertices of the dodecahedron. Consider how many ways there are of inscribing a regular tetrahedron in a dodecahedron. (There are 10. Recall that a tetrahedron can be inscribed in a cube in two ways, and a cube can be inscribed in the dodecahedron in five different positions.)

Why can't this technique of connecting non-adjacent vertices be used to inscribe a polyhedron in a tetrahedron, octahedron, or icosahedron? (The reason is that the faces of these solids are triangles, which have no non-adjacent vertices.)

3. Students should try to estimate this volume by eye. They may wish to estimate the volume of the space between the inscribed tetrahedron and the cube.

4. Before completing the models, you might want to obtain an estimate of volumes by using sand, rice, or similar material. Glue just one tab of the regular tetrahedron; fill it with your material, and pour it into the cube. Repeat until the cube is filled. (You should be able to fill the tetrahedron three times.) Then finish the models to show that the ratio of volumes is exactly 1/3.

5a. The area of one square face of this cuboctahedron is 1/2 the area of one square face of the cube.

6b. The area of one triangular face of the cuboctahedron is 1/4 the area of one triangular face of the octahedron.

Teacher Comments for Activity 7 (continued)

Discussion/Extension: Platonic and Archimedean solids are not the only ones that can be inscribed in a sphere. Ask students to find some other polyhedra that can be inscribed in a sphere. (For example, any pyramid or prism with a regular polygonal base will do.) Ask them to find a convex solid that cannot be inscribed in a sphere. (Example: a cube with a corner cut off.)

Truncation of polyhedra is closely related to inscribing polyhedra within polyhedra. In truncation, a plane slices through the edges that meet at a vertex of the polyhedron. See A. Holden, *Shapes, Space, and Symmetry*, page 40, for photographs of the various stages of truncation from cube to octahedron and from octahedron to cube. Several Archimedean solids are created during the stages of truncation. Other truncations are shown in Holden, pp. 41-45. See also Student Project 4 and Student Project 9.

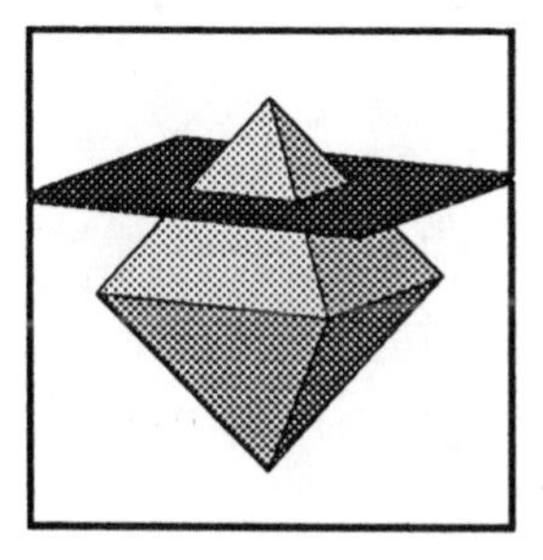

Activity 8: Cross Sections

1. Use the transparent cube. Wrap a rubber band *across* 4 edges of the cube, as shown. It does not have to bisect the edges of the cube, but it should be perpendicular to each edge of the cube it crosses.

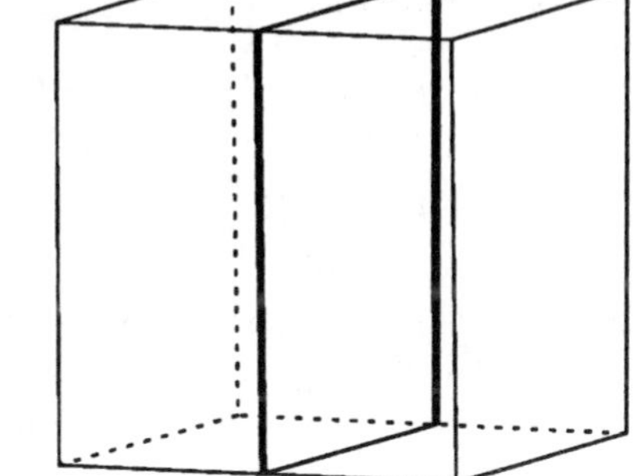

   a. How many faces of the cube are crossed by the rubber band?

      Four.

Imagine a plane slicing the polyhedron into two pieces. The cross section is the new face produced from the slice.

   b. What shape is outlined by the rubber band?

      A square.

Definition: A **cross section** of a polyhedron is the polygon produced when a plane intersects the polyhedron.

2. The rubber band in Exercise 1 represents a plane slicing through the cube, parallel to two faces of the cube. In this example, a square cross section is produced.

   a. Consider some other planes that form square cross sections when they intersect the cube. Are all square cross sections of the cube congruent?

      Yes.

3. Change the location of the rubber band so that it forms a cross section that is rectangular, but not square.

   a. Are all rectangular cross sections of the cube congruent?

      No.

   b. If you wanted to produce a rectangular cross section with the largest area possible, where should you put the rubber band? Sketch the rubber band on the picture.

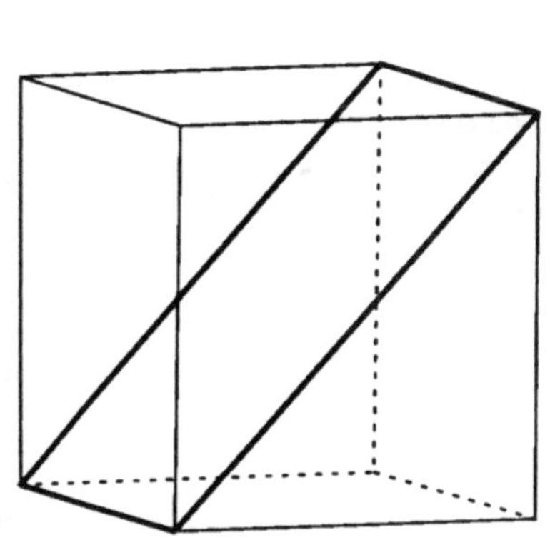

      Put the rubber band along the opposite edges of the cube.

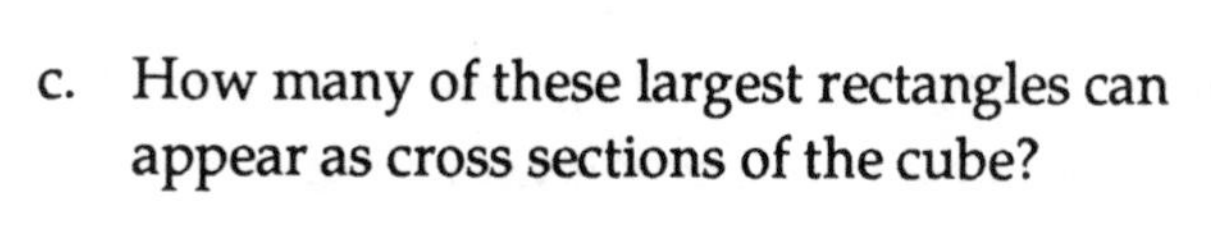

   c. How many of these largest rectangles can appear as cross sections of the cube?

      Six.

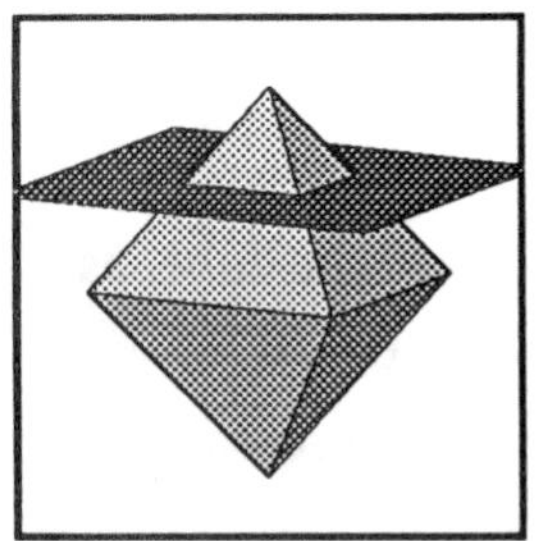

Activity 8 (continued)

4. Experiment with your cube and rubber band to find examples of other quadrilateral cross sections, such as a rhombus (but not a square) or a parallelogram (but not a rhombus or rectangle). Sketch your results, labeled appropriately, on the drawing of the cube.

   See Teacher Comments.

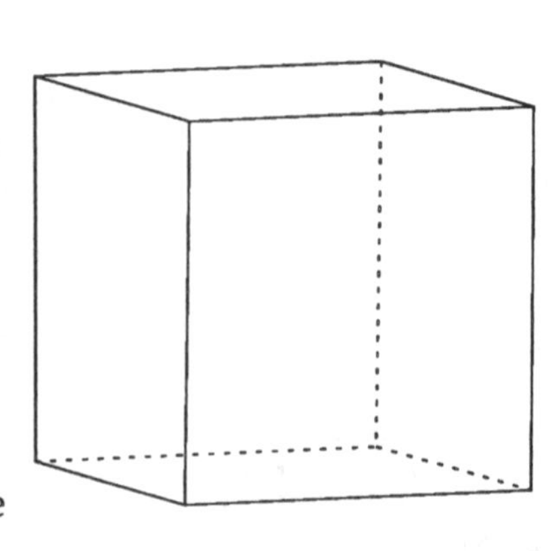

5. a. Describe how to slice the cube to form an equilateral triangle as a cross section of the cube. (Where are its vertices?)

   Slice off a corner of the cube. The vertices of the equilateral triangle should be equidistant from the vertex of the cube.

   b. Describe a largest equilateral triangle which can be obtained as a cross section of the cube. (Where are its vertices?)

   The vertices of a largest equilateral triangle will be adjacent to a single vertex of the cube.

   c. How many of these largest equilateral triangles can appear as cross sections of the cube?

   Eight.

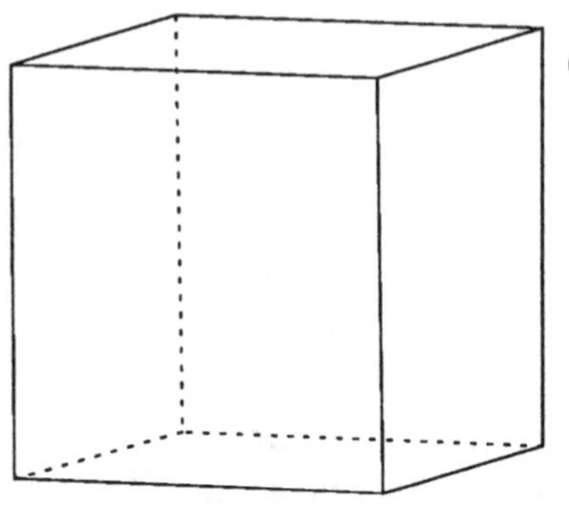

6. Experiment with your cube and rubber band to find examples of other triangular cross sections, such as an isosceles triangle or scalene triangle. Is it possible to get a right triangle or an obtuse triangle? *No.*

   Sketch your results, labeled appropriately, on the drawing to the left.

   See Teacher Comments.

Can you find a pentagonal cross section of the cube? Is it possible to find a cross section of the cube which is a regular pentagon?

7. a. How many faces of the cube must be cut by a plane to produce a hexagonal cross section?

   Six.

   b. Place a rubber band on your cube to demonstrate a cross section which is a regular hexagon. Sketch the rubber band on the picture.

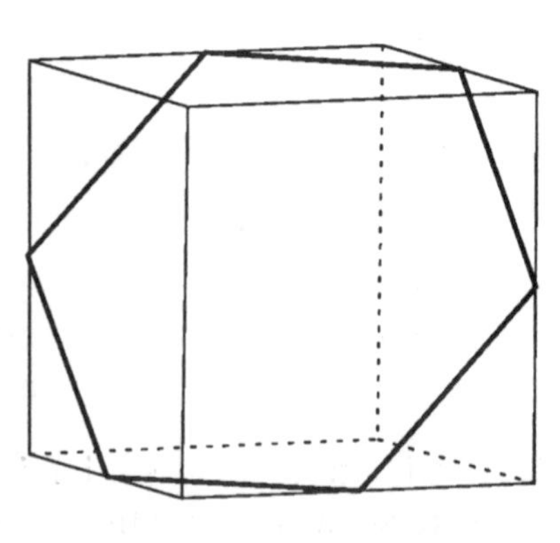

Teacher Comments for Activity 8

Rubber bands stretched around the transparent cube suggest polygonal cross sections. Cross sections are one of the most common and useful ways to describe three-dimensional objects and are extremely important in architecture and engineering. Here, students are directed to find square cross sections of the cube and then experiment to find what other shapes are possible as cross sections. They find the largest cross sections which are rectangles or equilateral triangles.

Background Knowledge: Polygons; quadrilaterals—rectangle, parallelogram, rhombus; triangles—acute, right, obtuse, scalene, isosceles; area of plane figures.

Presenting the Activity: Be sure students understand that a cross section of a three-dimensional object is planar. A plane may intersect a polyhedron at just a vertex, edge, or face, but these are not usually considered cross sections of the polyhedron. In each of the exercises in this activity, students must be sure that the stretched rubber band gives the outline of a planar polygon. Have students tilt the cube and sight the outline of the rubber band through it to check that it lies in a plane.

Note that the shape of cross sections varies with the position of the cut. You may wish to cut an apple or potato to demonstrate.

Materials: Transparent cube; rubber bands (size 14 or 16). Optional: non-permanent (water-based) marker; granular material such as birdseed, rice, or sand to pour into the cube.

Comments on Activity Questions

2a. Ask how these slicing planes are positioned with respect to the cube. There are three distinct orientations, corresponding to the top, front, and side of the cube in which a slicing plane is perpendicular to four faces and parallel to two faces of the cube. Square cross sections can also be formed by a plane that slices through the adjacent faces, cutting off a triangular prism.

3a. There are an infinite number of different-shaped rectangular cross sections, but only two distinct ways of slicing the cube to obtain them. Either (1) cut through two opposite (parallel) faces and a pair of faces that links them, or (2) cut through two adjacent faces and the pair of faces that links them. In both (1) and (2), two edges of the rectangle are parallel to four other edges of the cube that are not cut by that cross section. Students may discover that square cross sections can also be formed by a cut of type (2).

4. There are many distinct ways to cut the cube with a plane to get a quadrilateral cross section; even several different ways to get a rhombus or a parallelogram. In each case, the stretched rubber band representing the cross section must cross exactly four faces of the cube and it must outline a planar figure.

5. It is difficult to keep the rubber band in position to outline a triangular cross section; you may wish to have students outline these on the transparent cube using a non-permanent marker.

5c. You obtain a largest equilateral triangle when you slice off an entire corner and all the edges leading to it. There are eight corners on the cube, so there are eight equilateral triangles of largest area.

6. A triangular cross section is formed by cutting through three edges of the cube that meet at a vertex—just cut off the corner. By varying the angle of the cut, an infinite variety of triangles can be

formed. However, all angles in these triangles are acute.

7b. For a regular hexagon, the rubber band must cut through the midpoints of the edges it crosses.

It is possible to obtain a pentagonal cross section from a hexagonal one. Put one point of the stretched rubber band at a vertex of the cube so that exactly five faces are crossed by the band. However, it is not possible to obtain a regular pentagon as a cross section of the cube.

Discussion/Extension: This activity can be enhanced by pouring birdseed, rice, or other fine granular material in the cube. Fill it about 1/3 full to start; later, increase the amount to half. Close the cube tightly and secure it with some clear tape. Tilting the cube to various positions and letting the granular material settle creates cross sections—they are the flat top of the substance in the cube. Stretch a rubber band around the cube to outline the cross section or trace it on the cube with a non-permanent marker. Note the different shapes of cross sections that divide the volume of the cube in half. All these cross sections go through the center of the cube.

A challenging activity is to determine what cross sections are possible in the other Platonic solids. (See A. Holden, *Shapes, Space, and Symmetry*.) See Student Project 14 for a surprising cross section of the regular tetrahedron. Another challenging activity is to calculate and compare areas of largest cross sections of the cube of each type: triangle, rectangle, rhombus, hexagon.

Student Project 1: Semiregular Tessellations

Materials

Several triangles, squares, hexagons, octagons, and dodecagons (net pages SP-1A and SP-1B).

Background Knowledge

Activities P-1 and P-3.

In this project, you will determine how many different semiregular tessellations are possible.

Review

1. How do you find the measure of each interior angle of a regular polygon with n sides?
2. What's the difference between a *regular* tessellation and a *semiregular* tessellation?
3. You can arrange three triangles and two squares around a vertex with no gaps, since the total angle measure around the vertex will be $3(60°) + 2(90°) = 360°$. The code 3-3-3-4-4 can be used to describe the arrangement here (the numbers tell how many sides each shape has; they are ordered according to the arrangement of the shapes around the vertex). How else might three triangles and two squares be arranged around a vertex? (Use a code like the one above to describe this arrangement.)

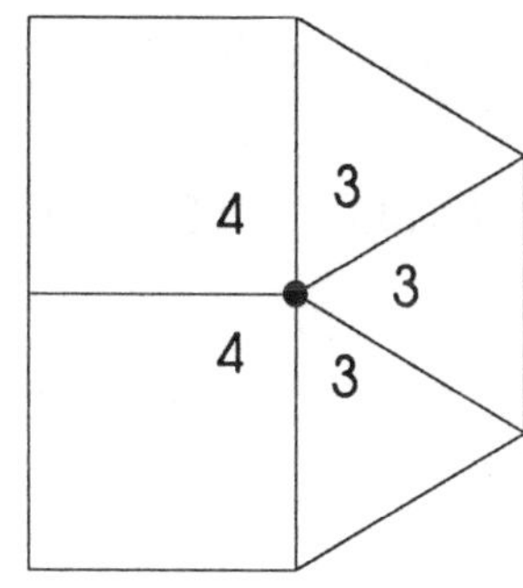

Experiment

4. Make several copies of the polygons on net pages SP-1A and SP-1B and cut them out. What arrangements of polygons (using more than one kind) can you fit around a vertex without any gaps? List as many as you can (you should find 12), using a code to describe them. (Be sure to consider different orders for each set of shapes.)

   3-3-3-4-4; 3-3-4-3-4; 3-6-3-6; 3-3-6-6; 3-3-3-3-6; 3-4-4-6; 3-4-6-4; 4-8-8; 4-6-12; 3-3-4-12; 3-12-12; 3-4-3-12

5. Now, for each arrangement in Exercise 4, try to continue the tessellation by adding more of the same kinds of polygons. Be sure that the arrangement of polygons around each vertex is the same. A few of them cannot be continued this way. See if you can find these and explain why they don't continue to form a semiregular tessellation. *3-3-6-6, 3-4-4-6, 3-3-4-12, and 3-4-3-12 don't work.*

6. Here are codes for some other arrangements of two or more regular polygons about a vertex. For each one, show that the total angle measure is 360°.

   a. 5-5-10 $2(108°) + 144° = 360°$

   b. 4-5-20 $90° + 108° + 162° = 360°$

   c. 3-7-42 $60° + 900°/7 + 7200°/42 = 360°$

   d. 3-8-24 $60° + 135° + 165° = 360°$

   e. 3-9-18 $60° + 140° + 160° = 360°$

   f. 3-10-15 $60° + 144° + 156° = 360°$

7. Without making all the regular polygons needed to try out the arrangements in Exercise 6 above, we can argue that none of these arrangements will form a semiregular tessellation. For example, both 6a and 6b involve a pentagon surrounded by two non-congruent polygons. That is, each vertex is surrounded by three angles—one measuring 108° (from the pentagon) and two others (a and b) which are not equal to each other. But that arrangement cannot be continued, as the following diagram illustrates.

Student Project 1 (continued)

Therefore, neither 6a nor 6b can be continued to form a semiregular tessellation.

Each of the remaining four cases in Exercise 6 has a triangle surrounded by two non-congruent polygons. Explain why no tessellation is possible in these cases.

Conclusion

8. How many different semiregular tessellations are there altogether?

Eight. They are: 3-3-3-4-4, 3-3-4-3-4, 3-6-3-6, 3-3-3-3-6, 3-4-6-4, 4-8-8, 4-6-12, and 3-12-12.

References

J. Britton and D. Seymour, *Introduction to Tessellations*, Chapter 3 and Appendix

K. Critchlow, *Order in Space*

B. Grünbaum and G. Shephard, *Tilings and Patterns an Introduction*, Chapter 2

P. O'Daffer and S. Clemens, *Geometry: An Investigative Approach*, Chapter 3

Student Project 2: Rotation Symmetries of the Regular Dodecahedron and the Regular Icosahedron

Materials

Regular dodecahedron and icosahedron from Activity 1, straws or pipe cleaners.

Background Knowledge

Activity 2.

Review

1. What does it mean for a polyhedron to have n-fold rotation symmetry?

Experiment

2. Pick up your dodecahedron and put a straw or pipe cleaner through a pair of opposite vertices. What kind (?-fold) of rotation symmetry does the dodecahedron have around this axis? *3-fold.*

   How many such axes does the dodecahedron have? *10*

3. Now put a straw or pipe cleaner through the centers of two opposite faces. What kind of rotation symmetry does this demonstrate? *5-fold.*

   How many such axes does the dodecahedron have? *Six.*

4. Finally, put a straw through the midpoints of two opposite edges, passing through the center of the polyhedron. What kind of rotation symmetry does this demonstrate? *2-fold.*

   How many such axes does the dodecahedron have? *15*

5. Repeat your investigations on the regular icosahedron. Summarize your results for the dodecahedron and the icosahedron in a table similar to the one in Activity 2, Exercise 8. Compare the rotation symmetries of the two shapes.

   The icosahedron has six 5-fold axes through opposite vertices, ten 3-fold axes through the centers of opposite faces, and fifteen 2-fold axes through the midpoints of opposite edges.

Student Project 3: Reflection Symmetry of the Platonic Solids

Materials

Seeds for the five Platonic solids (net pages SP-3A through SP-3E); three rectangular mirrors, at least 6 inches by 6 inches.

Background Knowledge

Activity 2.

Review

1. Describe the reflection planes of a cube.

Experiment

Lay one mirror flat, with the reflective side up. Place the other two mirrors upright, at a 45° angle to each other. You may want to make a tape hinge for the two upright mirrors.

2. Build the cube seed. Place it so that the "MIRROR" faces are flat against the two standing mirrors. You should see the image of a cube in the mirror kaleidoscope. Explain this image in terms of the reflection planes of the cube. How many images of the cube seed make the whole cube? *16*

3. Build the octahedron seed. Describe the seed itself. What shape is it? *An irregular tetrahedron.*

   Place the seed in the mirror kaleidoscope to see the image of an octahedron.

4. For the remaining three Platonic solids, remove the bottom mirror. Build the seeds, and place them against the standing mirrors. Close the mirrors until the seed is held in place. What is the measure of the angle between the two standing mirrors:

   a. for the tetrahedron seed? *60°*

   b. for the dodecahedron seed? *36°*

   c. for the icosahedron seed? *36°*

Challenge

The smallest seed for each Platonic solid is formed when all of the reflection planes cut the Platonic solid into parts. Describe the minimum seed for each Platonic solid, and determine how many of its images make the whole solid.

References

Films: *Dihedral Kaleidoscopes*

Symmetries of the Cube

Materials: *Root Blocks* by Rhombics (these blocks are the smallest cube seed)

Student Project 4: The Truncated Octahedron

Materials

8 hexagon panels, 6 square panels, 12 triangle panels (net page SP-4), rubber bands.

Background Knowledge

Activities 1 and 3.

Review

An Archimedean solid is built from two or more different kinds of regular polygons, so that each vertex of the solid is exactly the same. See the **Glossary** for pictures of the 13 Archimedean solids.

Definition

Truncation of a polyhedron occurs when all its corners or edges are sliced off in the same manner.

In this project, you will build a truncated octahedron and three corners of the original octahedron to demonstrate the truncation.

Experiment

Build the truncated octahedron by attaching hexagons and squares so that there are two hexagons and one square surrounding each vertex. Stand the figure on one of its square faces.

1. Build three pyramid-like corners, each with four triangles meeting at a vertex. Place one corner on the square face at the top of the truncated octahedron, and hold the other two on the square faces that show in the picture above.

   If you attached one of these corners to *each* square face of the truncated octahedron, what shape would you have? *A regular octahedron.*

2. A truncated octahedron is formed by slicing the corners off a regular octahedron so that the faces of the new solid formed are regular polygons. Demonstrate this now by removing the corners formed from triangles.

   What shape is the new face that is created when one corner is sliced off? *A square.*

3. The truncated octahedron is unique among the Archimedean solids in that several copies of this figure will pack together without any gaps.

   The angle between two faces of a polyhedron is called a **dihedral angle**. Look up the dihedral angles of the truncated octahedron in one of the references below, and show that there is a combination of these angles which has a sum equal to 360°. (This explains why the packing works.)

   $2(125°16') + 109°28' = 360°$

4. Each Platonic solid has a truncated form that is an Archimedean solid. For example, imagine truncating all the corners of a cube so that portions of the original edges remain and all the new faces are regular polygons.

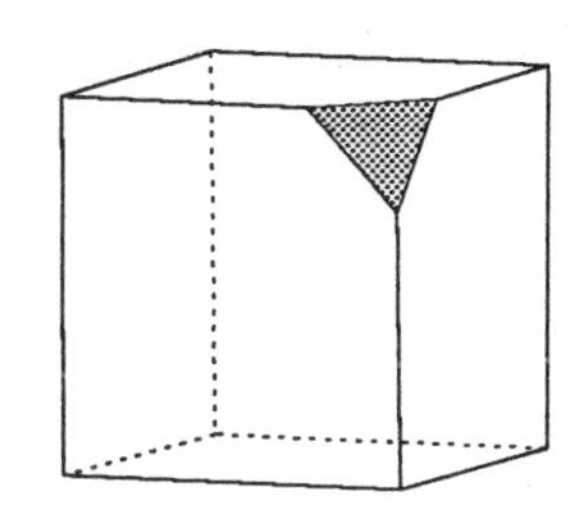

   What shape is the new face that is created when one corner is sliced off?

   A triangle.

Student Project 4 (continued)

After all the corners have been truncated, what shape replaces each original square face?

A regular octagon.

5. Repeat Exercise 4 for the tetrahedron, dodecahedron, and icosahedron. For each truncated solid, how is the number of new faces related to the original Platonic solid?

   It equals the number of faces plus vertices of the original Platonic solid, since each truncated corner produces a new face and each truncated face produces a new face.

   How is the shape of the new faces related to the original Platonic solid?

   Each face of the original Platonic solid is replaced by a face with twice as many edges, and each corner is replaced by a face that is a cross section of that corner.

For more on the truncated icosahedron, see Student Project 9.

References

K. Critchlow, *Order in Space* (has measures of dihedral angles)

H. M. Cundy and A. Rollet, *Mathematical Models* (has measures of dihedral angles)

A. Holden, *Shapes, Space, and Symmetry*

P. O'Daffer and S. Clemens, *Geometry: An Investigative Approach*, Chapter 4

H. Steinhaus, *Mathematical Snapshots*, Chapters 7 and 8

Student Project 5: The Rhombic Dodecahedron

Materials

Cube pattern (net page SP-5A), 6 pyramids (net page SP-5B), transparent cube, tape.

Background Knowledge

Activity 3.

Experiment

This project relates the rhombic dodecahedron to other shapes, particularly the cube.

Cut out the cube pattern. Glue the two halves together as indicated. Score the lines.

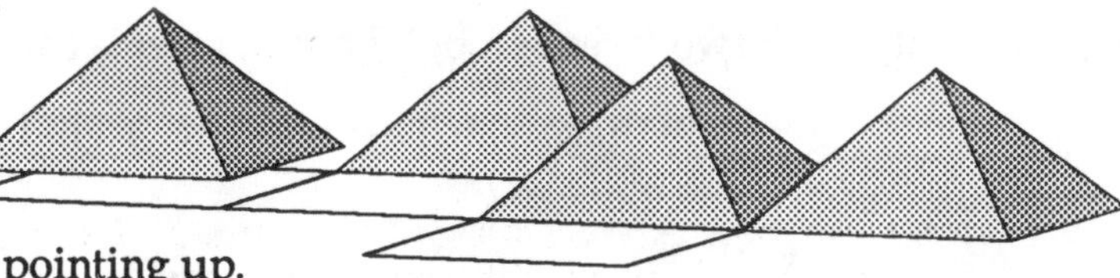

Build six pyramids, using net page SP-5B as a pattern. Glue one pyramid onto each square of the cube pattern, with the pyramids pointing up.

Fold the shape up to make a cube, with the pyramids on the inside. Your cube should be the same size as the transparent cube.

1. Now unfold the pyramids and wrap them around the transparent cube, with the base of a pyramid touching each face of the cube. What is the name of the solid formed on the outside? *A rhombic dodecahedron.*
2. Your transparent cube demonstrates a cube inscribed in a rhombic dodecahedron. How is each edge of the cube related to a face of the rhombic dodecahedron? *Each edge of the cube coincides with the short diagonal of a rhombus face.*
3. What is the volume of the rhombic dodecahedron compared to the volume of the cube inside? *It is twice as much.*
4. Suppose you draw the long diagonals of each face of the rhombic dodecahedron. These would form the edges of what inscribed shape? *An octahedron.*

References

Arthur Loeb, "The Rhombic Dodecahedron and Its Relation to the Cube and the Octahedron," *Shaping Space*, ed. M. Senechal and G. Fleck

Ian Stewart, "How to Succeed in Stacking," *New Scientist*, vol. 131, no. 1777, July 13, 1991

Student Project 6: Dihedral Angles

Materials

Scientific calculator, nets for the regular tetrahedron and regular octahedron (net page SP-6).

Background Knowledge

Activity 1, definition of trigonometric functions, Pythagorean Theorem, Law of Cosines.

In this project, you will calculate the dihedral angles of some Platonic solids. You need to know some trigonometry; a scientific calculator is recommended, as well.

Definitions

The angle at which two faces of a polyhedron meet is called a **dihedral angle**.

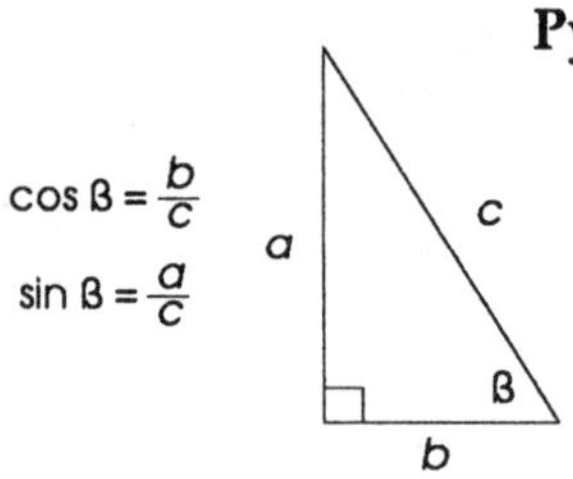

$\cos ß = \frac{b}{c}$

$\sin ß = \frac{a}{c}$

a c ß b

Pythagorean Theorem: In a right triangle with legs a and b and hypotenuse c, $a^2 + b^2 = c^2$.

In a right triangle, the **cosine** of an angle ß is defined to be the ratio of the length of the side adjacent to ß to the length of the hypotenuse. The **sine** of ß is the ratio of the length of the side opposite ß to the length of the hypotenuse.

arcfunctions: If $x = \cos ß$, then $ß = \arccos x$. If $y = \sin ß$, then $ß = \arcsin y$.

Law of Cosines: In any triangle with legs a, b, c and corresponding opposite angles A, B, and C, the following three rules apply:

$$a^2 = b^2 + c^2 - 2bc \cos A$$
$$b^2 = a^2 + c^2 - 2ac \cos B$$
$$c^2 = a^2 + b^2 - 2ab \cos C$$

Experiment

In a Platonic solid, all dihedral angles have the same measure. The measure of the dihedral angle is a property of the solid.

Build the octahedron and tetrahedron by folding along the solid lines and gluing the tabs.

1. What is the measure of the dihedral angle of a cube? *90°*

2. Now consider the octahedron and the tetrahedron. Hold them together by matching one pair of faces. Notice that each of the other faces of the tetrahedron is coplanar with a face of the octahedron. This shows us that the dihedral angles of the two solids are supplementary. Because of this property, octahedra and tetrahedra together will pack space with no gaps. The framework of edges in this packing (called an octet truss) forms a lightweight but rigid structure, and so it is often used in architecture and scaffolding.

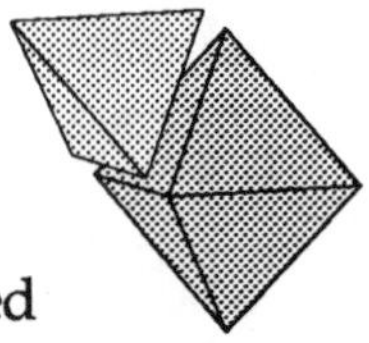

   Now we'll calculate the measures of these two angles.

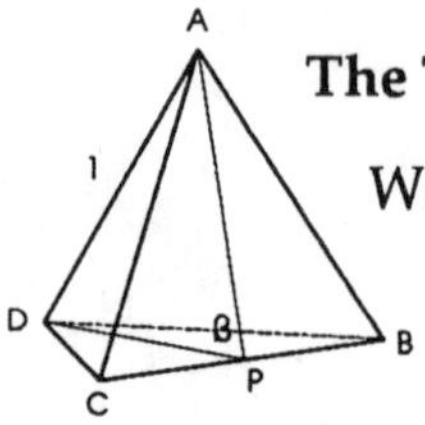

A 1 D ß B C P

The Tetrahedron

We are looking for the dihedral angle between faces ABC and BCD. First we need a cross-section plane perpendicular to $\overline{BC}$. We form this plane by drawing the altitude $\overline{AP}$ of triangle ABC and the altitude $\overline{DP}$ of triangle DCB. Angle APD is the dihedral angle ß.

Student Project 6 (continued)

To do the calculations, let's assume each edge of the tetrahedron is equal to 1.

3. Use the Pythagorean Theorem to show that $AP = \sqrt{3}/2 = DP$. *$AP^2 + (1/2)^2 = 1$, $AP^2 = 3/4$, $AP = \sqrt{3}/2$*

4. By the Law of Cosines, $AD^2 = AP^2 + DP^2 - 2(AP)(DP)\cos \beta$. Show that $\cos \beta = 1/3$.
   $1 = 3/4 + 3/4 - 2(\sqrt{3}/2)(\sqrt{3}/2)\cos \beta$; $3/2 \cos \beta = 1/2$; $\cos \beta = 1/3$

5. Use a table or calculator to find an approximate value for the measure of ß.
   $\beta \approx 70.5°$

The Octahedron

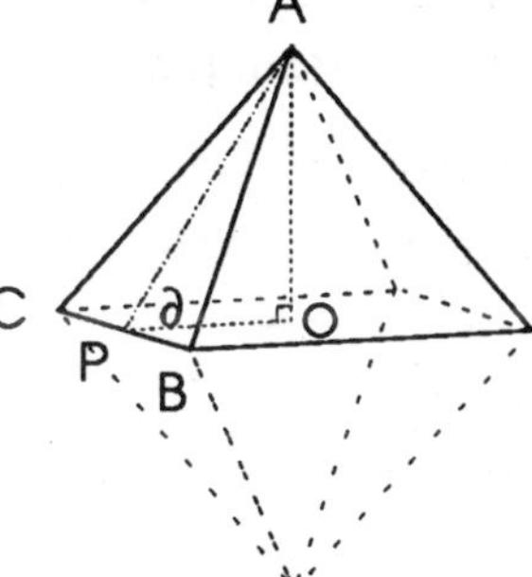

Think of the octahedron as built from two regular square pyramids. The dihedral angle ∂ between the base and one face of the pyramid equals half the dihedral angle of the octahedron. We find this angle ∂ by drawing the altitude $\overline{AO}$ of the *pyramid* and the altitude $\overline{AP}$ of triangle *ABC*. The dihedral angle ∂ is angle *APO*.

6. Assume the length of each edge of the square pyramid is equal to 1.
   Then $\overline{OP} = 1/2$ and $\overline{AP} = \sqrt{3}/2$. Why? *$\overline{OP}$ is half the length of each edge of the square. For $\overline{AP}$, see Excerise 3.*

7. Show that $\cos \partial = 1/\sqrt{3}$. *$\cos \partial = (1/2)/(\sqrt{3}/2) = 1/\sqrt{3}$*

8. Use a table or calculator to find an approximate value for the measure of ∂. *$\partial \approx 54.7°$*

9. The dihedral angle of a regular octahedron is 2∂. What is the measure of this angle? *$2\partial \approx 109.5°$*

Check that your results in Exercises 5 and 9 give supplementary angles.

Challenge

You can use trigonometric identities to prove that the sum of the dihedral angle of a tetrahedron and the dihedral angle of an octahedron is exactly 180°.

Begin with the results in Exercises 4 and 7: $\cos \beta = 1/3$, $\cos \partial = 1/\sqrt{3}$.

9. Use the identity $\cos \partial = \sqrt{(1 + \cos 2\partial)/2}$ to show that $\cos 2\partial = -1/3$.
   $1/\sqrt{3} = \sqrt{(1 + \cos 2\partial)/2}$; $1/3 = (1 + \cos 2\partial)/2$; $\cos 2\partial = -1/3$

   This means that 2∂ and ß are supplementary. Why? *2∂ and ß have opposite cosines. Since each angle is smaller than 180°, they must be supplementary.*

Dihedral Angles of Other Polyhedra

In an Archimedean solid, the dihedral angles need not have the same measure. In fact, the cuboctahedron and the icosidodecahedron are the only two Archimedean solids whose dihedral angles are all equal. (Why?)

11. Look up the dihedral angles of the truncated octahedron in a reference below, and show that there is a combination of these angles which has a sum equal to 360°. This shows that truncated octahedra will pack space with no gaps. *2(125°16') + 109°28' = 360°*

12. Archimedean duals have congruent faces and equal dihedral angles. Find an Archimedean dual that will pack space. *The rhombic dodecahedron (dihedral angle = 120°).*

Student Project 6 (continued)

References

For dihedral angles of regular polyhedra, see H. S. M. Coxeter, *Introduction to Geometry*, Chapter 10, Section 4. This contains a general method of computing the dihedral angle ß of any Platonic solid and derives the following general formula:

For a Platonic solid whose faces have n sides, with r faces meeting at each vertex,
$\beta = 2 \arcsin [\cos (\pi/r)/\sin (\pi/n)]$.

Another method of calculating dihedral angles for Platonic solids is in Chapter 4 of P. O'Daffer and S. Clemens, *Geometry: An Investigative Approach*. Packing space with regular and semiregular polyhedra is also discussed.

For calculated values of dihedral angles of Platonic solids and other polyhedra, see H. M. Cundy and A. Rollett, *Mathematical Models* or K. Critchlow, *Order in Space*.

For more on space packing and the octet truss (invented by Buckminster Fuller as the basis of "spaceframes"), see J. Kappraff, *Connections*, Chapter 10, and A. Edmundson, *A Fuller Explanation*.

The space packing of octahedra and tetrahedra is pictured in H. Steinhaus, *Mathematical Snapshots*, and A. Holden, *Shapes, Space, and Symmetry*.

Student Project 7: Applications of Euler's Formula

Materials

None.

Background Knowledge

Activity 5, algebra of inequalities.

Euler's Formula is extremely useful in analyzing convex polyhedra, as you'll see in this project.

Review

Euler's Formula ($f + v = e + 2$) is a relationship which relates the number of faces, vertices, and edges of any convex polyhedron.

Experiment

In Activity 5, you demonstrated that the relationships below hold for all polyhedra:

$$2e \geq 3f \qquad\qquad 2e \geq 3v$$

1. Combine these inequalities with Euler's Formula to show that every convex polyhedron has at least six edges. *$f + v = e + 2$; $3f + 3v = 3e + 6$; $2e + 2e \geq 3e + 6$; $e \geq 6$*

2. Show that every convex polyhedron has at least four vertices and at least four faces.
   $3f + 3v = 3e + 6$; $3f + 2e \geq 3e + 6$; $3f \geq e + 6 \geq 12$; $f \geq 4$. $2e + 3v \geq 3e + 6$; $3v \geq e + 6 \geq 12$; $v \geq 4$

3. Show that there is no convex polyhedron with exactly seven edges. (Hint: Suppose $e = 7$ for some convex polyhedron. Then use the inequalities above to determine restrictions on the number of faces and vertices the polyhedron might have. Show that this violates Euler's Formula.)

   Student Projects 9 and 11 have other applications of Euler's Formula. Still more may be found in the references below. *If $e = 7$, then $3f \leq 2e = 14$; $f \leq 4$. Similarly, $3v \leq 14$; $v \leq 4$. Then $f + v = 8 \neq e + 2 = 9$. Now use the result in exercise two to conclude that $f = v = 4$.*

References

A. Beck, M. Bleicher, and D. Crowe, *Excursions Into Mathematics*, Chapter 1

J. Pedersen and P. Hilton, *Build Your Own Polyhedra*, Section 12.3

H. Rademacher and O. Toeplitz, *The Enjoyment of Mathematics*, Chapters 12 and 13

Student Project 8: Deltahedra

Materials

58 triangle panels, rubber bands.

Background Knowledge

Activity 1, Activity 5.

Definition

A **deltahedron** is a convex polyhedron all of whose faces are congruent equilateral triangles.

Aside from the Platonic solids, the deltahedra are the only convex polyhedra having all faces congruent regular polygons.

In this project, you will use Euler's Formula ($f + v = e + 2$) to determine how many deltahedra can be built.

Experiment

You already know three deltahedra—the regular tetrahedron, the regular octahedron, and the regular icosahedron. But more are possible, since not all vertices have to be alike.

Let v_3 be the number of vertices at which 3 faces meet.

Let v_4 be the number of vertices at which 4 faces meet.

Let v_5 be the number of vertices at which 5 faces meet.

(Why is there no need for a variable v_6?)

Explain the reasoning behind the following equations:

1. $v = v_3 + v_4 + v_5$

   This counts all the vertices.

2. $3v_3 + 4v_4 + 5v_5 = 2e$

   This counts the edges twice.

3. $3v_3 + 4v_4 + 5v_5 = 3f$

   Each face has 3 vertices.

4. Now combine these three equations with Euler's Formula to show that the following relationship must hold: $3v_3 + 2v_4 + v_5 = 12$

   $f + v = e + 2;\ 6f + 6v = 6e + 12;$

   $6v_3 + 8v_4 + 10v_5 + 6v_3 + 6v_4 + 6v_5 = 9v_3 + 12v_4 + 15v_5 + 12;$

   $3v_3 + 2v_4 + v_5 = 12$

This equation has 19 solutions. Some of these are indicated in the chart on the next page.

5. Fill in the rest of the chart with values of v_3, v_4, and v_5 that satisfy the equation.

Although there are 19 solutions to the equation in Exercise 4, only eight of these deltahedra can actually be built. (This was proved in 1947 by H. Freudenthal and B. L. van der Waerden.)

6. Try to build the models shown in your chart, using triangles and rubber bands from Activity 1. You will need 90 triangles (58 if you don't try to make the Platonic ones) to make all eight deltahedra. Try to explain why some of the cases don't produce deltahedra (remember that they must be convex).

Student Project 8 (continued)

Finally, for the models which can be built, write their names in the chart. Use the following list, which tells how many faces each deltahedron has:

Regular Tetrahedron (4 faces)
Triangular Dipyramid (6 faces)
Regular Octahedron (8 faces)
Pentagonal Dipyramid (10 faces)
Snub Disphenoid (also called a Siamese dodecahedron) (12 faces)
Triaugmented Triangular Prism (14 faces)
Gyroelongated Square Pyramid (16 faces)
Regular Icosahedron (20 faces)

v_3	v_4	v_5	Name
0	0	12	Regular Icosahedron
0	1	10	
0	2	*8*	*Gyroelongated Square Pyramid*
0	3	*6*	*Triaugmented Triangular Prism*
0	4	*4*	*Snub Disphenoid*
0	5	*2*	*Pentagonal Dipyramid*
0	6	*0*	*Regular Octahedron*
1	0	*9*	
1	1	*7*	
1	2	*5*	
1	3	*3*	
1	4	*1*	
2	0	*6*	
2	1	*4*	
2	2	*2*	
2	3	*0*	*Triangular Dipyramid*
3	0	*3*	
3	1	*1*	
4	0	0	Regular Tetrahedron

References

A. Beck, M. Bleicher, and D. Crowe, *Excursions into Mathematics,* Chapter 1

J. Pedersen and P. Hilton, *Build Your Own Polyhedra,* pp. 77-80

B. M. Stewart, *Adventures Among the Toroids,* Chapter IV

Models: *Geometric Playthings,* J. and K. Pedersen

Student Project 9: The Truncated Icosahedron

Materials

12 pentagon panels, 20 hexagon panels, rubber bands.

Background Knowledge

Activity 1, Activity 4, Activity 5.

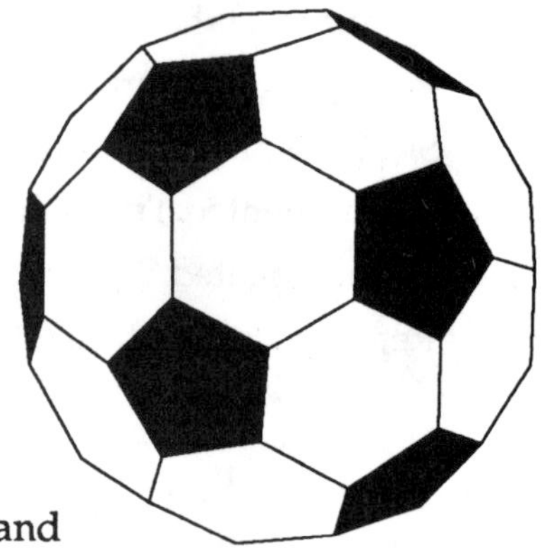

A standard soccer ball is formed from sewn leather polygons that are regular hexagons and pentagons. The soccer ball is made exactly like the Archimedean solid known as the **truncated icosahedron.**

The truncated icosahedron is the only possible semiregular polyhedron with regular hexagons and regular pentagons as faces. In this project, you will show why this is true.

Review

1. a. First, explain why any polyhedron built from regular hexagons and regular pentagons must have exactly three faces meeting at each vertex.

   b. Using part a, explain why the equation $2e = 3v$ must hold for any such polyhedron.

Experiment

2. Use Euler's Formula ($f + v = e + 2$) and the fact that $2e = 3v$ to show that any convex polyhedron built from regular hexagons and pentagons must have exactly 12 pentagonal faces. (Hint: Let x be the number of pentagonal faces, y the number of hexagonal faces. Find expressions for f, v, and e in terms of x and y. Substitute these into Euler's Formula, and solve for x.)

   $f = x + y;\ 5x + 6y = 2e;\ 5x + 6y = 3v;\ f + v = e + 2;\ 6f + 6v = 6e + 12;\ 6x + 6y + 10x + 12y = 15x + 18y + 12;\ x = 12$

3. The only possible vertex configurations for a semiregular polyhedron built from regular hexagons and regular pentagons are (i) two hexagons and a pentagon, or (ii) two pentagons and a hexagon. Explain why the second option is impossible.

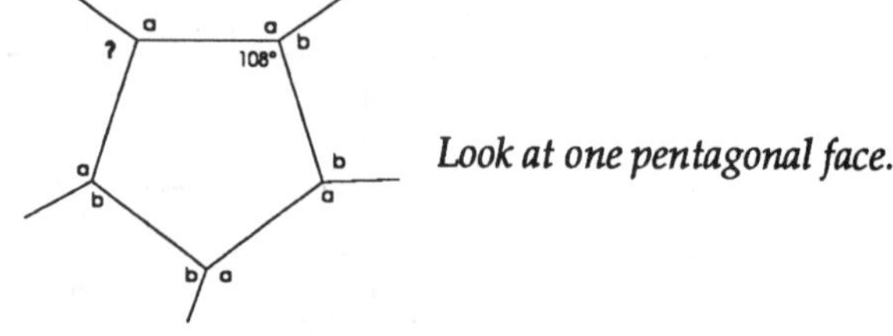

Look at one pentagonal face.

4. So far, you have shown that every vertex is surrounded by two hexagons and a pentagon, and that there are exactly 12 pentagon faces. Show that there must be exactly 20 hexagon faces. (Hint: every pentagon must be surrounded by five hexagons.)

   Each pentagon face borders five hexagons; each hexagon touches three different pentagons. Thus, there are (12 x 5)/3 = 20 hexagon faces.

5. Build a truncated icosahedron, using 12 pentagon panels and 20 hexagon panels.

References

A. Beck, M. Bleicher, and D. Crowe, *Excursions Into Mathematics*, Chapter 1

R. Curl and R. Smalley, "Fullerenes," *Scientific American*, Oct. '91

Student Project 10: The Toroid

Materials

15 square panels, 6 triangle panels, rubber bands.

Background Knowledge

Activity 1, Activity 5.

Experiment

The polygon panels and rubber bands used in Activity 1 can be used to make a shape with a hole in it—an example of a **toroid**.

First, form the inside hole: Arrange 3 squares and 3 triangles as in the pattern at the right. Attach the end triangle to the end square to close the loop.

Now, attach three squares together in a row.

Then attach them to the squares on your loop. Close the top loop, also.

So far you have made the shape below.

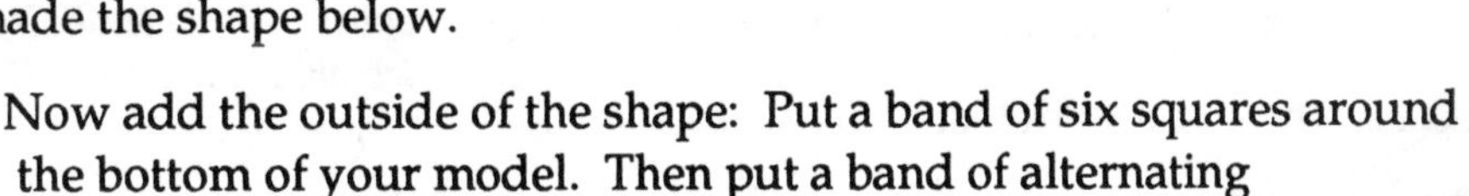

Now add the outside of the shape: Put a band of six squares around the bottom of your model. Then put a band of alternating triangles and squares around the top.

This completes the toroid. An interesting property of this shape is that it does not satisfy Euler's Formula ($f + v = e + 2$). Demonstrate this by counting the number of vertices, edges, and faces of the finished shape. $f = 21;\ v = 18;\ e = 39$

What *is* the relationship between f, v, and e for this shape? $f + v = e$

Reference

B. M. Stewart, *Adventures Among the Toroids*, Chapter V

Student Project 11: Descartes' Theorem

Materials

None.

Background Knowledge

Activity P-1, Activity 5.

Review

The Exterior Angle Theorem says that for any convex polygon, the sum of the exterior angles is 360°.

Definition

Descartes' Theorem is the analogous statement about convex polyhedra. Instead of exterior angles, Descartes looked at the **defect** (or **deficit**) at each vertex of a convex polyhedron. To compute the defect at a vertex, sum all the face angles that meet at that vertex, and subtract the sum from 360°. The defect at a vertex gives a measure of how close the polyhedron is to being spherical.

Experiment

1. a. For each of the Platonic solids, find the defect at a vertex. *tetrahedron: 180°; octahedron: 120°; cube: 90°; dodecahedron: 36°; icosahedron: 60°.*

   b. Which solid has the smallest defect? *The dodecahedron.*

   c. Which solid seems to be the most round? *The dodecahedron.*

2. a. Find the sum of the defects for each Platonic solid. (Since the defect is the same at each vertex, you can just multiply the number of vertices of the solid by the defect at one vertex.) *tetrahedron: 720°; cube: 720°; octahedron: 720°; dodecahedron: 720°; icosahedron: 720°.*

   b. State a conjecture based on your findings: *The sum of the defects of any Platonic solid is 720°.*

To prove your conjecture, use the following variables:

f = number of faces
v = number of vertices
e = number of edges
n = number of sides on each face of the polyhedron
A = sum of all of the face angles of the polyhedron
d = defect at each vertex
D = sum of the defects

The key technique in the proof is to use two different ways to express A in terms of the other variables.

3. Assume you have a Platonic solid.

   a. First, we'll find an expression for A by adding all the interior angles of each polygon face.

   How many faces are there? f

   What is the sum of the interior angles of each face? $180°(n-2)$

   Complete: $A = 180° f(n-2)$

   b. Now, we'll find A in terms of the defect. Explain why the following equation must hold: $A = v(360° - d)$. *$360° - d$ gives the sum of the face angles at one vertex. Multiply by the number of vertices to obtain the sum of all the face angles of the polyhedron.*

   c. You have two equations involving A. Use the substitution principle to eliminate A. What equation does this give? $180° f(n-2) = v(360° - d)$

Student Project 11 (continued)

d. Now, use Euler's Formula ($v - e + f = 2$) and the facts that $fn = 2e$ (why?) and $vd = D$ (why?), to show that $D = 720°$. *$180°fn - 360°f = 360°v - vd$; $360°e - 360°f = 360°v - D$; $D = 360°(v - e + f) = 720°$*

4. a. Use the drawings of each of the Archimedean solids in the appendix to calculate the defect at each vertex and the sum of the defects for each Archimedean solid.

Name	Defect at Each Vertex	Sum of the Defects
Truncated Tetrahedron	*60°*	*720°*
Truncated Cube	*30°*	*720°*
Truncated Octahedron	*30°*	*720°*
Truncated Dodecahedron	*12°*	*720°*
Truncated Icosahedron	*12°*	*720°*
Cuboctahedron	*60°*	*720°*
Icosidodecahedron	*24°*	*720°*
Snub Cuboctahedron	*30°*	*720°*
Snub Icosidodecahedron	*12°*	*720°*
Greater Rhombicuboctahedron	*15°*	*720°*
Rhombicuboctahedron	*30°*	*720°*
Greater Rhombicosidodecahedron	*6°*	*720°*
Rhombicosidodecahedron	*12°*	*720°*

b. Does your conjecture in Exercise 2b apply to the Archimedean solids? *Yes.*

Challenge

René Descartes (French mathematician, 1596-1650) proved that *for any convex polyhedron*, the sum of all the defects at its vertices is exactly 720°. A proof by mathematician George Pólya (1887-1985) is similar to the one you did in Exercise 3 for regular polyhedra. It is complicated by the fact that not all the faces are the same, so we let f_3 denote the number of three-sided faces, f_4 be the number of four-sided faces, and so on. Similarly, not all the vertices are the same, so we let a_n be the sum of the face angles about the vertex v_n. Try the proof yourself, or see the references.

References

J. Pedersen and P. Hilton, *Build Your Own Polyhedra*, pp. 159-161

D. Schattschneider, *Counting It Twice*

Student Project 12: Dual Polyhedra

Materials

Cube, six pyramids (net page SP-12).

Background Knowledge

Activities 3 and 6.

Review

The cube and the regular octahedron are dual polyhedra; that is, the faces of the cube correspond to the vertices of the octahedron, and the vertices of the cube correspond to the faces of the octahedron.

1. How does the number of *edges* of a cube compare to the number of *edges* of an octahedron?

Experiment

In this project, you will build a model showing a cube and an octahedron interpenetrated so that each edge of one of them is the perpendicular bisector of an edge of the other.

Build the cube first by folding along the solid lines and gluing the tabs. Copy the pyramid pattern on net page SP-12 so that you have six pyramid nets.

Color the triangular faces of the six pyramids all the same color, then build the pyramids. Glue the square face of each pyramid onto a face of the cube, as indicated by the dotted lines. The colored pyramids show the corners of an octahedron as though it interpenetrates the cube. (The videotape *The Platonic Solids* shows this as one view of the dual cube and octahedron.)

2. The two interpenetrating polyhedra share a shape as their inner core. What is the shape formed by the intersection of the cube and the octahedron? *A cuboctahedron.*

3. Each edge of the cube is crossed by an edge of the octahedron; these are called **dual edges**. Each pair of dual edges forms the diagonals of what two-dimensional figure? *A rhombus.*

4. How many *pairs* of dual edges are there? *12*

5. You should be convinced, from Exercises 3 and 4, that the smallest convex shape that will contain your model is a rhombic dodecahedron. How is this solid related to the inner solid in Exercise 2? *It is the dual.*

6. Other pairs of dual polyhedra can be arranged in this way (see A. Holden, *Shapes, Space, and Symmetry*, p. 9). In particular, a tetrahedron interpenetrated with its dual (another tetrahedron) forms a shape called the **stella octangula**, which is pictured here.

   What polyhedron is the inner core of the stella octangula? *An octahedron.*

   What is the smallest convex shape that will hold a stella octangula? *A cube.*

References

A. Holden, *Shapes, Space, and Symmetry*

Visual Geometry Project, *The Stella Octangula*

Visual Geometry Project, *The Platonic Solids*, Videotape

M. Wenninger, *Dual Models*

Student Project 13: The Volume of the Octahedron Inscribed in a Cube

Materials

Transparent cube, octahedron, 4 small tetrahedra, 4 "cube pieces" (net page SP-6 and 13).

Background Knowledge

Activities 6 and 7; volume of a pyramid.

Review

A regular octahedron can be inscribed in a cube so that each vertex of the octahedron touches the center of a face of the cube. This demonstrates the dual relationship of the octahedron and the cube.

Experiment

1. Build the octahedron (net page SP-6 and 13) and place it in the transparent cube. Estimate: How do you think the volume of the octahedron compares to the volume of the cube? *Answers will vary.*

In this project, you will find the exact ratio for these volumes. To do this, you must build several small pieces to place in the cube with the octahedron.

2. First, build four cube pieces. Show how to assemble these four pieces (but don't glue them together) to make an octahedron congruent to the one in the cube. How does the volume of one cube piece compare to the volume of the octahedron? *It is 1/4 the volume of the octahedron.*
3. Build one small tetrahedron (net page SP-6 and 13). Place the tetrahedron and one cube piece side by side, with the cube piece resting on its equilateral triangle face.

   Explain why the tetrahedron and the cube piece have the same volume. *They have the same height and the same base.*
   How does the volume of the tetrahedron compare to the volume of the octahedron? *It is 1/4 the volume of the octahedron.*
4. Build three more small tetrahedra (net page SP-6 and 13). Glue the four small tetrahedra onto the faces of the octahedron to form a large tetrahedron. Show how the large tetrahedron fits in the transparent cube.

   How does the volume of the octahedron compare to the volume of the large tetrahedron? *It is 1/2 the volume of the large tetrahedron.*
5. In Activity 7, you found that the volume of a tetrahedron inscribed in a cube is equal to 1/3 the volume of the cube. Using this fact, explain how the volume of the inscribed *octahedron* compares to the volume of the cube. *It is 1/6 the volume of the cube.*
6. (*optional*) You can make a cube puzzle by building 12 cube pieces, 8 small tetrahedra, and an octahedron. Challenge a friend to fit all these pieces into the cube.

References

M. Laycock and M. Smart, *Create a Cube*

Visual Geometry Project, *The Stella Octangula*

Student Project 14: The Tetrahedron Puzzle

Materials

Tetrahedron puzzle net (net page SP-14).

Background knowledge

Activity 8.

Experiment

1. Make two copies of the net, and build two identical models. Try to fit the two shapes together to form a regular tetrahedron. Describe the shape of the two faces that need to be matched to form the tetrahedron. *The square faces need to be matched.*
2. Describe precisely how to cut a tetrahedron into these two congruent pieces. *See first part of below.*
3. Challenge a friend to make the tetrahedron from the two shapes. What makes this puzzle so tricky is that the cross section plane that forms the two parts of the tetrahedron is not a reflection plane. Explain what you have to do to put the two shapes together to form a tetrahedron.

Think of the tetrahedron as if it were inscribed in a cube. Slice the cube in half, bisecting four parallel edges. You will be slicing through the midpoints of four edges of the tetrahedron.

Put the two square faces together as if the square face were a reflection plane, then twist one of the shapes 90°.

References

P. O'Daffer and S. Clemens, *Geometry: An Investigative Approach*, pp. 125-126 has three dissection puzzles.

M. Smart and M. Laycock, *Create a Cube*, has several dissection puzzles of the cube.

D. Stonerod, *Puzzles in Space*, has several dissection puzzles of the Platonic solids.

S.T. Coffin, *The Puzzling World of Polyhedral Dissections*, has several dissection puzzles of the Platonic solids.

References

Abelson, H. and di Sessa, A., *Turtle Geometry: The Computer as a Medium for Exploring Mathematics*, Cambridge, MA: MIT Press, 1981.

Mathematical explorations using LOGO are well suited to the subjects of polygons and symmetry.

Alper, Joseph, "Archimedes, Plato Make Millions for Big Oil," *Science* 248 (June 8, 1990) pp. 1190-1191.

By making polyhedral models, an Exxon Chemist solved the structure of boggsite, whose chemical properties make it valuable as a catalyst in the manufacture of gasoline and other substances.

Banchoff, Thomas, *Beyond the Third Dimension: Geometry, Computer Graphics and Higher Dimensions*, New York: W. H. Freeman, 1990.

Beautiful color illustrations accompany a discussion of geometric shapes in 2, 3, and even 4 dimensions. Cross-sections and projections of the regular polyhedra are demonstrated.

Beard, Col. R.S., *Patterns in Space*, Palo Alto: Creative Publications, 1973.

This compendium of photos and nets of the Platonic solids and other polyhedra also tabulates numerical relations between edges and other linear dimensions for many polygons and polyhedra.

Beck, A., Bleicher, M., and Crowe, D., *Excursions into Mathematics*, New York: Worth Publishers, 1969.

Chapter 1 is devoted to polyhedra and contains a proof of Euler's formula as well as several applications. Other sections cover the regular polyhedra, deltahedra, solids with tunnels, and the *n*-dimensional cube. Kepler's drawings of the Platonic solids and his model for the orbits of the planets based on the Platonic solids are reproduced.

Blackwell, W., A. I. A, *Geometry in Architecture*, Berkeley: Key Curriculum Press, 1984.

An architect discusses and richly illustrates, in both photographs and original drawings, the many applications of polygons and polyhedra to architecture and design.

Burckhardt, J. J. et al, "A Tribute to Leonard Euler," *Mathematics Magazine* 56 (1989) pp. 262-307.

Seven authors review the life, work, and times of this great mathematician.

Coffin, Stewart T., *The Puzzling World of Polyhedral Dissections*, New York: Oxford University Press, 1990.

This guide has a vast array of polyhedral puzzles, with directions for constructing your own.

Coxeter, H. S. M., *Introduction to Geometry*, 2nd ed., New York: John Wiley & Sons, 1969.

Chapter 10 is devoted to the Platonic solids: their nets, projections, Euler's formula, determination of dihedral angles, and dual polyhedra.

Coxeter, H. S. M., "Kepler and Mathematics," *Vistas in Astronomy*, Vol 18, A. Bear and P. Bear, eds., New York: Pergamon, 1975.

Coxeter discusses Kepler's infatuation with the Platonic solids and some of his discoveries. He gives details of Kepler's (false) model of the orbits of the planets based on Platonic solids and notes Kepler's own later comments on this.

Critchlow, Keith, *Order in Space*, New York: The Viking Press, 1969.

This book has many drawings of regular and semi-regular polyhedra, shows their inter-relationships, contains tables of measurements and information on how some can pack space.

Cundy, H. M., and Rollett, A. P., *Mathematical Models*, New York: Oxford Press, 1961.

This book has nets and pictures of the Platonic and Archimedean solids and their duals, and many stellated polyhedra. It contains vital statistics: numbers of vertices, faces, edges, measures of dihedral angles, ratios of edge lengths, etc. It contains a drawing of five cubes inscribed in a dodecahedron and five tetrahedra in a dodecahedron. Section 3.7.14 on isomerism indicates how four of the Archimedean solids can be "wrongly assembled."

References (continued)

Curl, Robert, and Smalley, Richard, "Fullerenes," *Scientific American* (Oct. '91), vol. 265, no. 4., pp. 54-63.

Discusses the properties of "Bucky Balls," different variations of a newly created carbon form called Buckminsterfullerene. This molecule, shaped like a truncated icosahedron, is a third form of pure carbon (the other two being diamond and graphite) and is creating a stir in the scientific community for the properties that stem from its icosahedral symmetry.

Edmondson, A., *A Fuller Explanation: The Synergetic Geometry of Buckminster Fuller*, Boston: Birkhauser, 1987.

One of the primary structural elements in Fuller's geometry is the "octet" truss built from the skeleton of packed regular tetrahedra and octahedra.

Euclid, *Elements*, c. 300 B.C.

Books XI and XII discuss polyhedra; Book XIII shows that there are only five regular polyhedra.

Federico, P. J., *Descartes on Polyhedra*, New York: Springer-Verlag, 1982.

This book gives historic details of Descartes' work on polyhedra, including his theorem on the sum of the defects of a convex polyhedron. Descartes' work is then compared with Euler's work on polyhedra a century later.

Franco, B. and Shimizu-Yost, J., *Using Tomoku Fusè's Unit Origami in the Classroom*, Key Curriculum Press, 1991.

Unit origami gives students hands-on experience folding three-dimensional shapes out of paper. This classroom guide prepares teachers and students for the more complex paper folds described in *Unit Origami*, by Tokomo Fusè. It offers an introduction to origami, easy folding activities, math questions to students about their folds, and practical tips for using *Unit Origami* in the classroom.

Galloway, John, "Nature's Second Favourite Structure,"*New Scientist* (31 March 1988), pp. 38-39.

According to this article, the helix is nature's favorite structure, but the pentagonal dodecahedron runs a close second. The article shows many illustrations from nature, particulary viruses, that have this symmetry.

Gardner, Martin, *The New Ambidextrous Universe*, New York: W. H. Freeman, 1990.

This very readable account of the variety of occurences and implications of reflection symmetry has a wealth of diagrams and pictures.

Gardner, Martin, *Time Travel and Other Mathematical Bewilderments*, New York: W. H. Freeman, 1988.

Chapter 13 discusses possibilities for tiling the plane with convex polygons (the pentagon problem is not yet completely solved). Chapter 14 poses questions on tiling by polyominoes and such.

Grünbaum, B. and Shephard, G., *Tilings and Patterns, an Introduction*. New York: W. H. Freeman, 1989.

This has accurate information on tilings, including historical background, corrections to errors in the literature (for example, about tilings by regular polygons), and an excellent bibliography.

Jacobs, Harold R., *Mathematics A Human Endeavor*, New York: W. H. Freeman, 1970.

Chapter 5 , "Symmetry and Regular Figures", has examples of tessellations with regular polygons, shows regular polyhedra and some possible truncations, and has diagrams showing polyhedra with their duals inscribed.

Holden, Alan, *Space, Shapes, and Symmetry*, New York: Columbia University Press, 1971.

This book has excellent photographs of regular, semiregular, and many other polyhedra fashioned from cardboard and/or wire. It gives clear visual depictions of the interrelationships of polyhedra, their symmetry, and the processes of truncation, duality, and packing.

References (continued)

Hume, Andrew, *Exact Descriptions of Regular and Semi-regular Polyhedra and Their Duals*, Murray Hill, NJ: AT&T Bell Laboratories, Computing Science Technical Report No. 130, 1986.

This 32 page booklet contains accurate nets and measurements (including dihedral angles) for the Platonic and Archimedean solids and their duals.

Kappraff, Jay, *Connections: The Geometric Bridge Between Art and Science*, New York: McGraw-Hill, 1991.

A text based on material developed for students of technology, this is a rich resource of ideas. It has chapters on tilings with polygons, the Platonic solids, transformations of the Platonic solids, space-filling by polyhedra, isometries and mirrors, and symmetry of the plane.

Kepler, J., *Harmonices Mundi*, Linz, 1619.

Kepler's investigations of the Platonic solids and other polyhedra are described here.

Kim, Scott, *Inversions*, New York: W.H. Freeman, 1989. (Reprint of original, Byte Books, 1981.)

This catalogue of "calligraphic cartwheels" has clever examples of half-turn and mirror symmetry in the written names of famous people, of slogans, and even concepts. An accompanying disk of computer software, *Letterforms and Illusions*, encourages users to see the multiple use of letters through their symmetry and to create original calligraphic symmetry puzzles.

Klaasen, Daniel, "Regular Polyhedra" in *Historical Topics for the Mathematics Classroom*, pp. 220-221, Washington, D. C.: National Council of Teachers of Mathematics, 1969.

This capsule discusses interest in the regular polyhedra chronologically from the Pythagoreans (c. 500 B. C.) to Schläfli (1814-1895). It contains a proof that there are only five regular polyhedra.

Lakatos, Imre, *Proofs and Refutations: The Logic of Mathematical Discovery*, New York: Cambridge University Press, 1976.

What is the definition of a polyhedron? This dialogue clearly shows the difficulties of how to say what we mean and to capture the essential characteristics. Attempts to define a polyhedron show flaws under scrutiny and several refinements are formulated in light of Euler's Theorem.

Loeb, Arthur L., "The Rhombic Dodecahedron and its Relation to the Cube and the Octahedron", in *Shaping Space*, ed. M. Senechal and G. Fleck, pp. 61-63 (see below).

This gives two different recipes for making a rhombic dodecahedron; both have the same volume.

McLean, K. Robin, "Dungeons, dragons and dice," *The Mathematical Gazette*, 74 (1990) pp. 243 - 256.

This articles addresses the properties of fair polyhedral dice.

Miyazaki, Koji, *An Adventure in Multidimensional Space, The Art and Geometry of Polygons, Polyhedra, and Polytopes*, New York: John Wiley and Son, 1986.

Beautiful color plates dramatically show a great variety of polyhedra and polyhedral packings.

O'Daffer, P. and Clemens, S., *Geometry: An Investigative Approach*, Menlo Park: Addison Wesley, 1976.

An excellent pre-service and reference text, this includes chapters on polygons (their symmetries, construction, theorems), tessellations by polygons (including an analysis of the semiregular ones), and polyhedra (regular and semiregular, symmetry, duals, space packing). Many activities and clear diagrams, especially for duality and space packing.

Pearce, Peter and Pearce, Susan, *Polyhedra Primer*, Palo Alto: Dale Seymour Publications, (Reprint of original, Van Nostrand Reinhold, 1978).

This book tabulates information on polygons, tessellations, polyhedra, dual polyhedra, space fillings, open packings and constructions. (Beware—not all of the definitions and drawings in this book are accurate.)

References (continued)

Pearce, Peter, *Structure in Nature is a Strategy for Design*, Cambridge, MA: M. I. T. Press, 1978.

This book covers various shapes that occur in nature and adapts them to effective structural design.

Pedersen, Jean and Hilton, Peter, *Build Your Own Polyhedra*, Palo Alto: Addison-Wesley, 1988.

By folding and braiding paper strips, a variety of polyhedra are constructed. Properties of polyhedra, as well as number theory that arises from the paper folding are discussed. Chapter 12 compares the volume of the cube with its inscribed tetrahedron and octahedron, rotation symmetries, and Euler's and Descartes' Theorems.

Plato, *Timaeus*, c. 360 B.C.

In this dialogue, Plato discusses the elements of air, fire, water, and earth and associates to each a regular polyhedron. This is the origin of the name "The Platonic solids."

Pugh, Anthony, *Polyhedra, A Visual Approach*, Berkeley: U. of California Press, 1976.

This book gives an overview and tabulates information on and relationships between the regular and semiregular polyhedra and their duals. It describes the 8 convex deltahedra and the 92 convex polyhedra with regular polygon faces. An appendix shows calculations of various measures.

Schattschneider, Doris, "Counting it Twice," *The College Mathematics Journal* (May 1991) and Agate 3 (1989) pp. 6-11.

Illustrations of the power of counting sets of objects twice are capped with an example of Pólya's proof of Descartes' Theorem for convex polyhedra.

Schattschneider, Doris, "Instruction in Geometry", *Mathematics Education in Secondary Schools and Two-Year Colleges: A Sourcebook*, Campbell, P., Grinstein, L., eds., pp. 123-165, N. Y.: Garland, 1988.

An essay and extensive annotated bibliography supply information on books, models, model-making kits, student activities, displays, and other resources for teaching geometry. Several sections focus on polygons and polyhedra.

Schattschneider, Doris, "Tiling the Plane with Congruent Pentagons," *Mathematics Magazine* 51 (1978) pp. 29-44, and 58 No. 5 (1985) cover and p. 308.

This gives a complete description of all types of convex pentagons that are known to tile the plane, including many equilateral ones, and the history of the pentagon tiling problem.

Schattschneider, Doris and Walker, Wallace, *M. C. Escher Kaleidocycles*, Petaluma, CA: Pomegranate Publications, 1987.

Seventeen ready to assemble polyhedral models (including the Platonic solids) are continuously covered with M. C. Escher's designs, accompanied by the story of the geometry behind them.

Senechal, Marjorie, and Fleck, George, *Shaping Space, A Polyhedral Approach*, Boston: Birkhäuser, 1988.

A rich collection of essays and activities on polyhedra (with lots of photos of polyhedra as they occur in nature, architecture, and art). These are a "must" for context, history, and applications. Key articles: "A Visit to the Polyhedron Kingdom," by M. Senechal, "Regular and Semiregular Polyhedra," by H. S. M. Coxeter, "Milestones in the History of Polyhedra," by Joe Malkevitch. The book's bibliography is excellent.

Serra, Michael, *Discovering Geometry, An Inductive Approach*, Berkeley: Key Curriculum Press, 1989.

A geometry text rich in exploratory activities. Chapter 7, "Transformations and Tessellations," has an excellent collection of Escher-type tessellations created by high school geometry students. A computer activity describes tessellating using LOGO.

References (continued)

Seymour, Dale and Britton, Jill, *Introduction to Tessellations*, Palo Alto: Dale Seymour Publications, 1989.

This introductory book addresses many geometric aspects of tessellations with polygons: symmetry, duality, derivation of all possible vertex configurations of regular polygons, the 8 semiregular and several demiregular tessellations. Recipes for Escher-like tessellations are also given. Accompanied by *Tessellation Teaching Masters.*

Smart, Margaret and Laycock, Mary, *Create a Cube*, Hayward, CA: Activity Resources, 1985.

A collection of activities in which various polyhedra are built and assembled to completely fill a cube box. (There are blackline masters for 26 different polyhedra nets.)

Steen, Lynn Arthur, ed., *On the Shoulders of Giants: New Approaches to Numeracy*, Washington, D.C.: National Academy Press, 1990.

Included in this collection of essays by mathematicians and mathematics educators are two important articles on geometry and its introduction at all levels: "Shape," by Majorie Senechal, and "Dimension," by Thomas F. Banchoff.

Stein, Sherman K., *Mathematics, the Man-Made Universe*, San Francisco: W. H. Freeman, 1976.

Chapters 5 and 6 discuss tiling and the role it plays in electrical currents and addresses the possibility of tiling a rectangle with other rectangles and squares. Chapter 15 shows a method of counting used in map coloring problems and derives Euler's formula.

Steinhaus, Hugo, *Mathematical Snapshots*, 3rd Ed., New York: Oxford University Press, 1983.

Chapters 7 and 8 have good photos to illustrate polyhedral space packing, the Platonic solids with their duals inscribed in them, cross sections of polyhedra, and other curiosities.

Stewart, B., *Adventures Among the Toroids: A Study of Orientable Polyhedra with Regular Faces*, 2nd ed., 1980. From the author: 4494 Wausau Rd., Okemos, MI 48864

A fascinating compendium of information about polyhedra with regular polygon faces, including many that have tunnels.

Stewart, Ian, "How to Succeed in Stacking," *New Scientist*, vol. 131, no. 1777, July 13, 1991.

The age-old problem of how to most efficiently stack spheres has finally been solved. A mathematician at the University of California at Berkeley proved recently that grocers have been stacking oranges the right way all along. The stacking method is related to the ways the octahedron and rhombic dodecahedron pack space.

Stonerod, David, *Puzzles in Space*, Hayward, CA: Activity Resources, 1983.

A collection of activities and puzzles with simple polyhedra, for which blackline masters are given.

Visual Geometry Project, *The Stella Octangula*, Berkeley, CA: Key Curriculum Press, 1991.

An activity book and accompanying videotape explore the many properties of this star-polyhedron and related Platonic solids. A manipulative kit is also available to facilitate model building.

Wenninger, M. J., *Polyhedron Models*, Also, *Dual Models*, London: Cambridge Univ. Press, 1971 and 1983.

These books contain directions and pictures of nets to build models of polyhedra from paper, including Platonic and Archimedean solids and their duals and many stellations and compounds.

Williams, Robert, *The Geometrical Foundation of Natural Structure*, New York: Dover Publications, 1979.

The geometry of the polyhedra in this book is derived from such disciplines as mathematics, chemistry, botany, physics, and virology. Chapter 3 has specific information about the regular polyhedra, i.e., surface area, volume, dihedral angle measure.

References (continued)

Model-Making Kits

Many model-making kits are available commercially; those suggested here are convenient for school orders from various catalog distributors of supplemental teaching materials for mathematics.

The Platonic Solids Manipulative Kit, Visual Geometry Project (Key Curriculum Press), includes polygon panels and everything else students need to complete these activities.

Cut and Assemble 3-d Geometrical Shapes, A. G. Smith (Dover Publications), includes materials for constructing 10 full color geometric solids.

Geometric Playthings to Color, Cut and Fold, Jean and Kent Pedersen (Dale Seymour Publications), contains models of the eight deltahedra and some flexagons.

Make shapes, G. Jenkins and A. Wild (Tarquin Publications), has a wide variety of models.

Sturdy plastic polygons such as *Polydrons and Googolplex* can be joined by snap-together hinges; these are excellent to investigate tessellations and build polyhedra.

"Straw" model-making kits such as Geo D-Stix can be used to make skeletal models.

Root Blocks (Rhombics) are fold-up-and-make models of the irregular tetrahedra that result when a cube is cut by its nine reflection planes. Many polyhedra can be built from these.

Symmetrics has kits of sturdy plastic panels that students can glue together to make many polyhedra.

Films and Videotapes

The Platonic Solids, Visual Geometry Project (Key Curriculum Press). A 15-minute computer animation illustrates the concepts students investigate in these activities. A 17-minute video for teachers shows how these materials can be used in the classroom.

Dihedral Kaleidoscopes (Coxeter) (1966; 13 min.; *Math Teacher* Jan. '73, p. 51) Presents basic ideas of symmetry by exhibiting the dihedral groups by means of reflection in two intersecting mirrors.

Symmetries of the Cube (Coxeter, Moser) (1971; 13 1/2 min.; *Math. Teacher* Dec. '72, p. 733) Shows symmetries of the cube and octahedron via reflections in the three mirrors of a dihedral kaleidoscope. Relates the construction of these kaleidoscopes to the cube and the octahedron.

Archimedean and Archimedean Dual Polyhedra and *Regular Convex Polyhedra: Why Exactly Five?* (Cal State Northridge, Lorraine Foster). The second video shows computer animation and students doing hands-on work to demonstrate why there are exactly five Platonic solids.

Many films are available from: International Film Bureau, 322 South Michigan Ave., Chicago, IL 60604.

Supplementary Materials

Pattern Blocks are colorful polygon blocks that can be used to make tessellations.

Mira is a rectangle of ruby plexiglas that has the reflective quality of a mirror, as well as a transparent quality. It is useful for investigating mirror symmetry and can be used as a construction tool.

Polyhedral dice illustrate the principle of "fairness," dependent on the uniformity of the shape.

Wooden or plastic models of the Platonic solids and other polyhedra.

Computer Software

The Geometer's Sketchpad (Visual Geometry Project) is a dynamic program for the Macintosh that can be used to construct regular polygons and tilings and to investigate symmetry in two-dimensions.

IQ-Plus Platonic Solids is a computer game that challenges you to match Platonic solids with varieties of their flat unfolded patterns.

Letterforms and Illusions (see Kim, Scott, above).

BLACKLINE MASTERS: Terminology, Glossary, Student Activities, Student Projects, and Nets

Terminology

Here are the most common terms used to describe two- and three- dimensional figures.

Polygons

Polygons are composed of connected line segments, called **edges,** which enclose a single region of the plane. Edges are also called the "sides" of the polygon. A **vertex** of a polygon is a point where exactly two edges meet, forming an **interior angle.**

vertex *edge* *interior angle*

The names of polygons describe how many edges they have:

triangle: three edges
quadrilateral: four edges
pentagon: five edges
hexagon: six edges
septagon: seven edges
octagon: eight edges
***n*–gon**: *n* edges

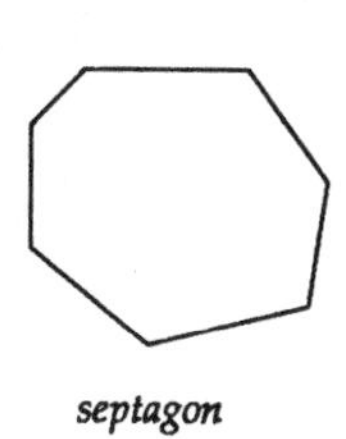

septagon

A polygon is **regular** if all of its edges have equal length and all of its interior angles have equal measure. A regular triangle is called an **equilateral triangle,** and a regular quadrilateral is called a **square.**

regular

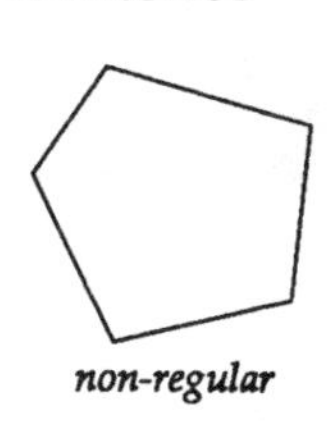

non-regular

A polygon is **convex** if any two points on its edges can be connected by a line segment which lies entirely inside the polygon. Otherwise, the polygon is **non–convex.**

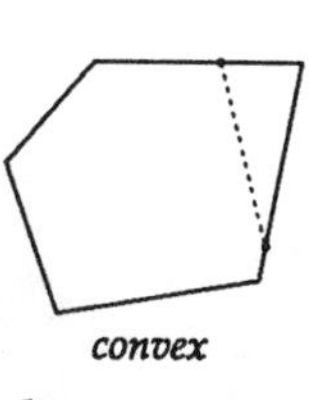

convex

non-convex

Polyhedra

Polyhedra are composed of polygons, called **faces,** which enclose a single region of space. An **edge** of the polyhedron is formed where exactly two faces are joined. A **vertex** of a polyhedron is a point where three or more edges meet.

vertex *face* *edge*

The names of polyhedra describe how many faces they have:

tetrahedron: four faces
hexahedron: six faces
octahedron: eight faces
***n*–hedron**: *n* faces

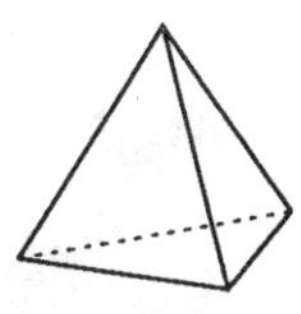

regular tetrahedron

A **pyramid** is formed when all vertices of a polygon are joined by line segments to a single point not in the same plane.

pentagonal pyramid

A polyhedron is **regular** if all of its faces are congruent regular polygons, and the same number of faces meet at each vertex in exactly the same way. A regular hexahedron is called a **cube.**

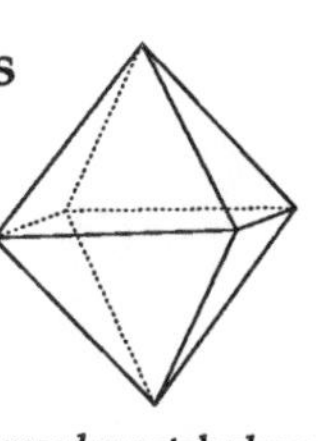

regular octahedron

A polyhedron is **convex** if any two points on its surface can be connected by a line segment which lies entirely inside or on the polyhedron. Otherwise it is **non–convex.**

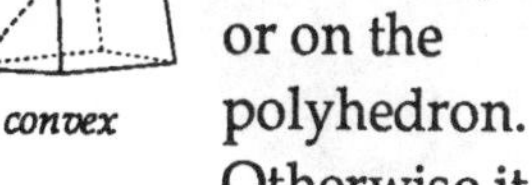

convex

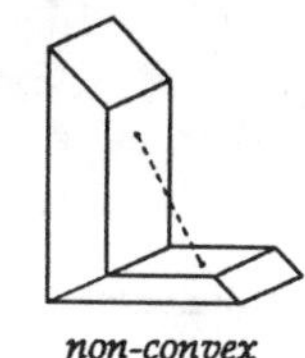

non-convex

Glossary

The Platonic Solids

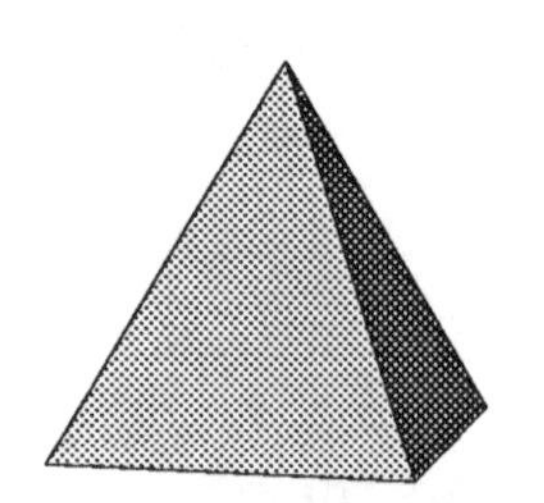
Tetrahedron

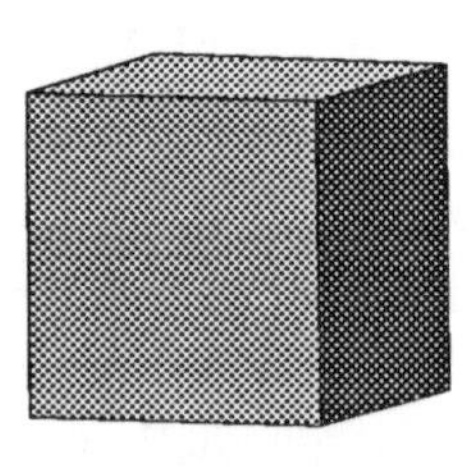
Cube

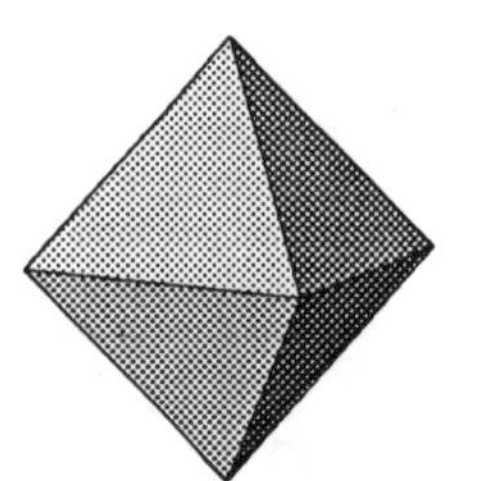
Octahedron

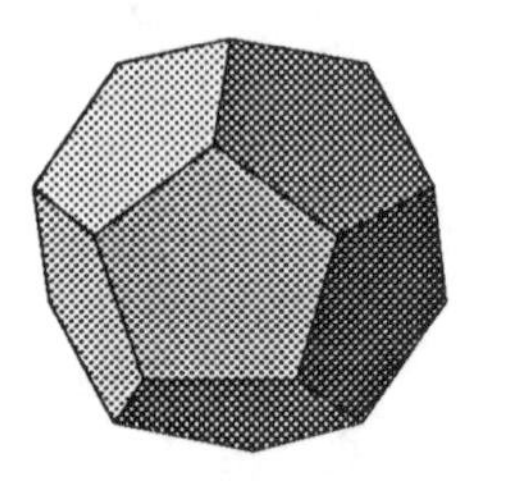
Dodecahedron

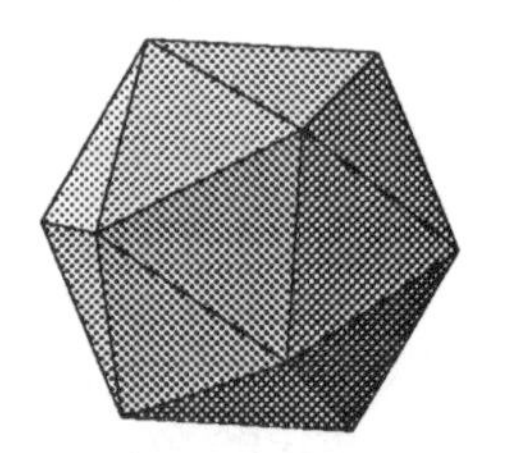
Icosahedron

The Archimedean Solids

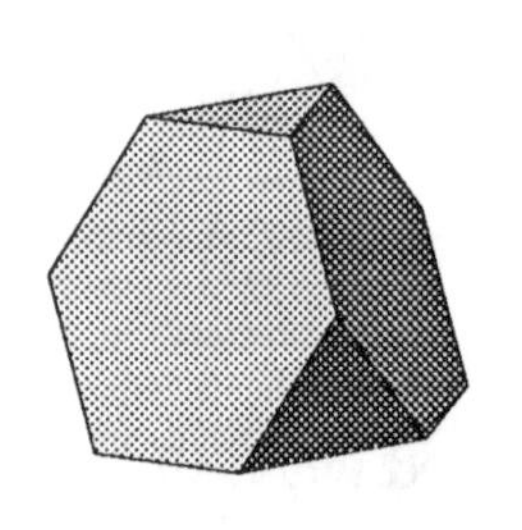
Truncated tetrahedron

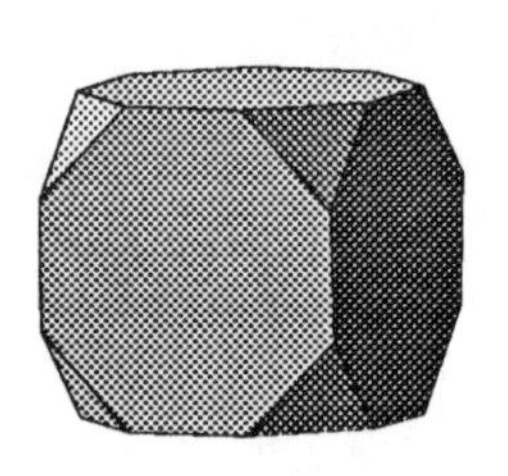
Truncated cube

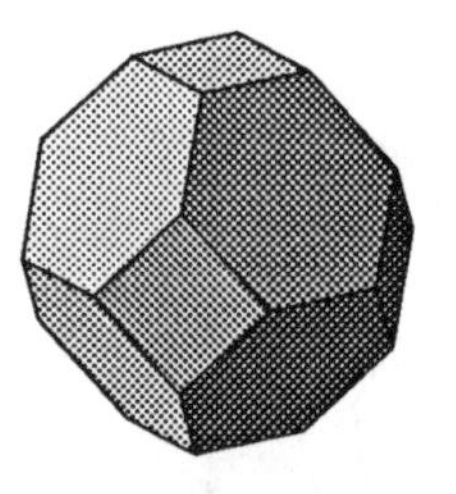
Truncated octahedron

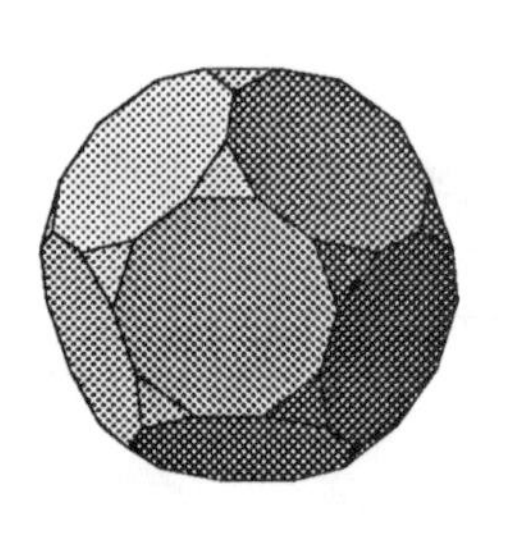
Truncated dodecahedron

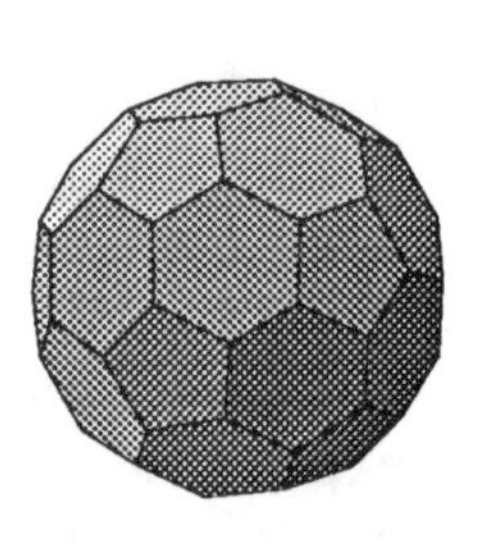
Truncated icosahedron

Cuboctahedron

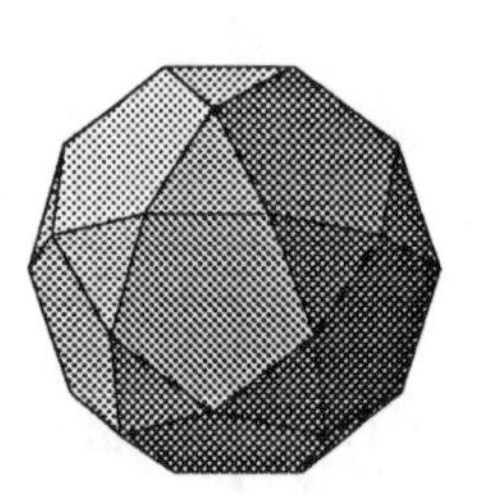
Icosidodecahedron

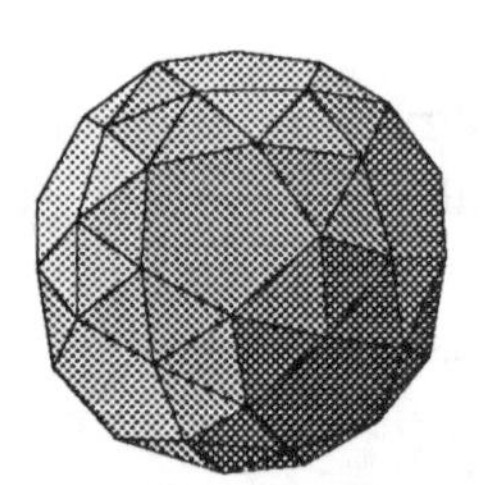
Snub dodecahedron

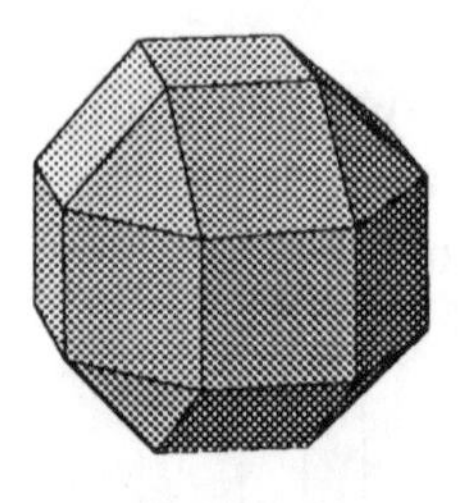
Rhombicuboctahedron

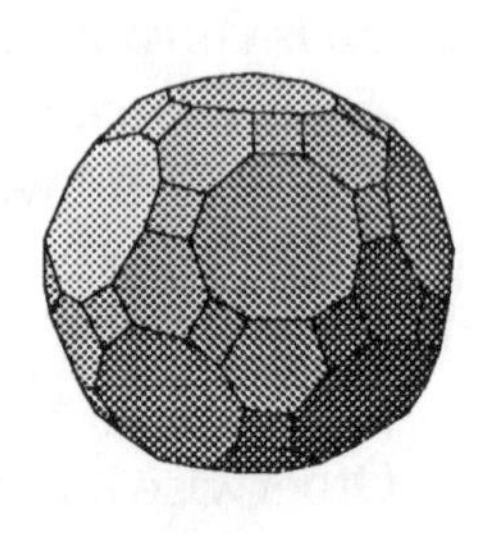
Great rhombicosidodecahedron

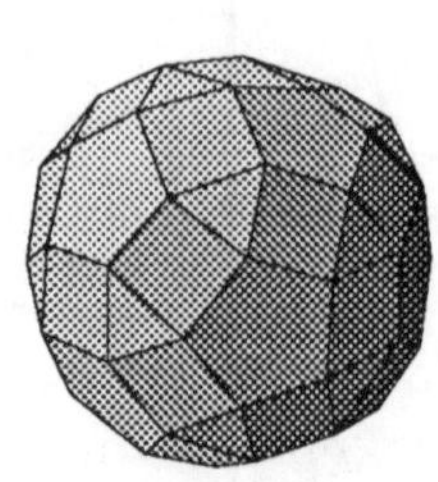
Rhombicosidodecahedron

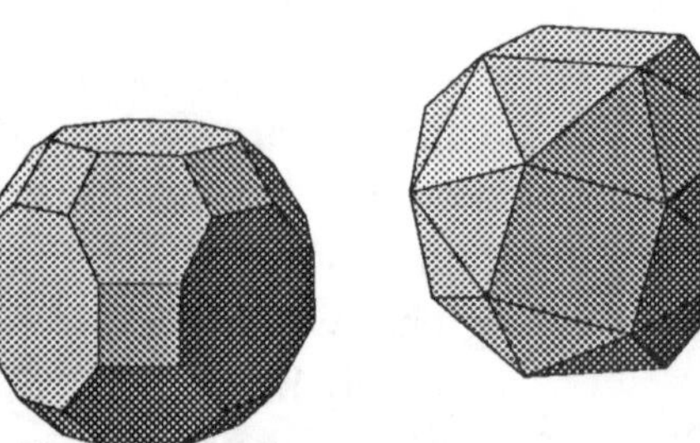
Great rhombicuboctahedron

Snub cube

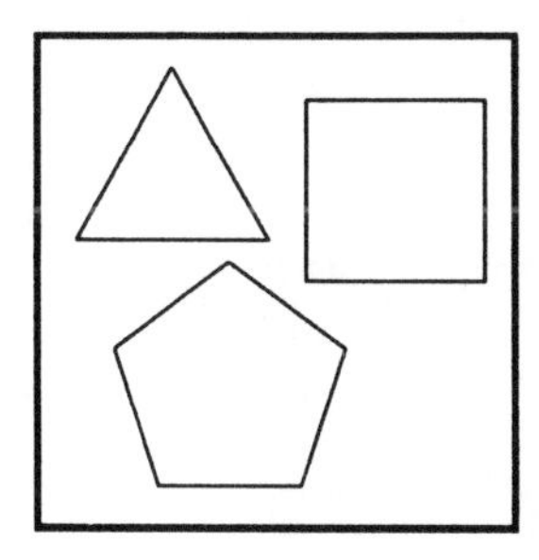

Activity P-1: Regular Polygons

In a **regular polygon**, all sides have the same length and all angles have the same measure.

Here are two equilateral triangles. Although one is larger than the other, their angle measures are the same.

1. What is the measure of each interior angle?

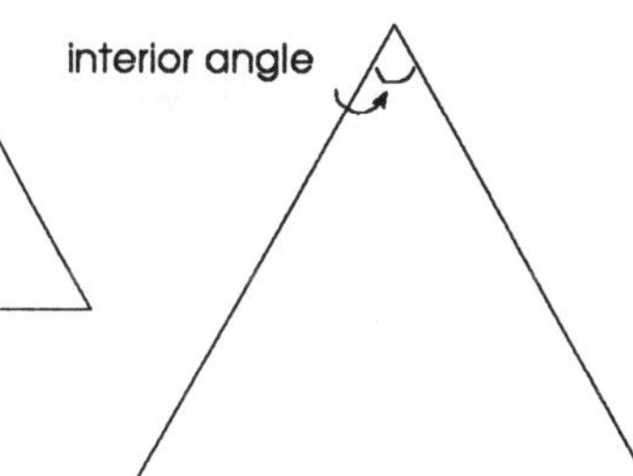

Explain how you arrived at your answer.

A triangle whose sides are all equal in length is called an ***equilateral triangle****. It is also a regular triangle, for all its angles have equal measure.*

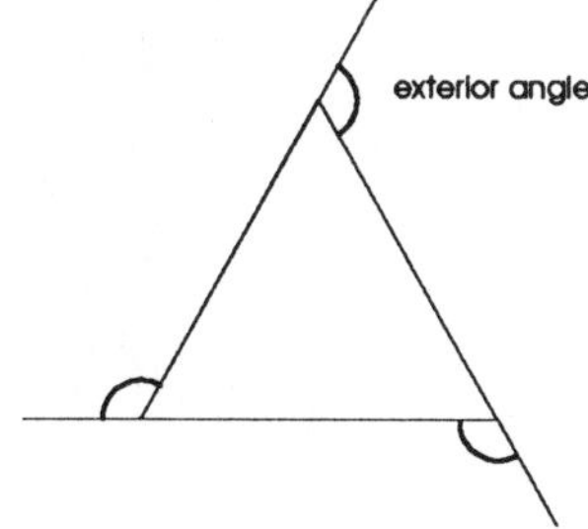

2. What is the measure of each exterior angle of any equilateral triangle?

Explain how you arrived at your answer.

A polygon with more than three sides, such as a rectangle, may have angles that are equal in measure but sides that are not equal in length.

A ***rhombus*** *is an example of a polygon whose sides are equal in length but whose angles may not be equal in measure.*

3. What is the sum of the measures of the three exterior angles of any equilateral triangle?

A regular polygon is both equilateral and equiangular.

Here's a way to check your answer in Exercise 3. Place your pencil flat on the paper as shown in the diagram. Now, rotate your pencil clockwise about its point, turning it through the exterior angle at A, until it lies against side $\overline{AB}$ of the triangle. Slide your pencil along the line through $\overline{AB}$ until the pencil point lies on vertex B. Repeat this procedure—rotate and slide along the side of the triangle—two more times. Your pencil should be back in its original position. Notice that your pencil made a complete turn of 360°.

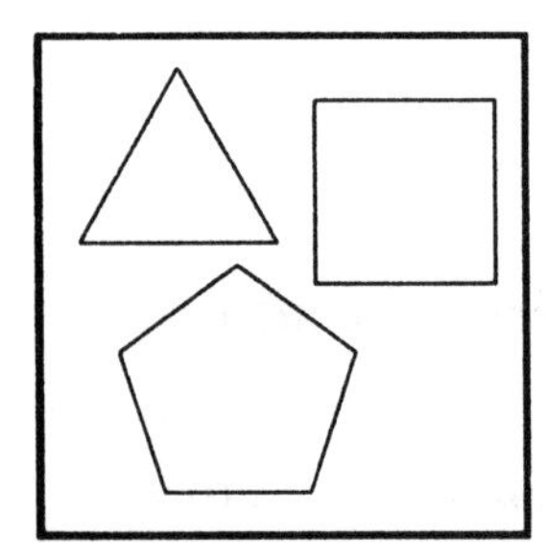

Activity P-1 (continued)

4. Use this pencil method to determine the sum of the exterior angles of the following regular polygons:

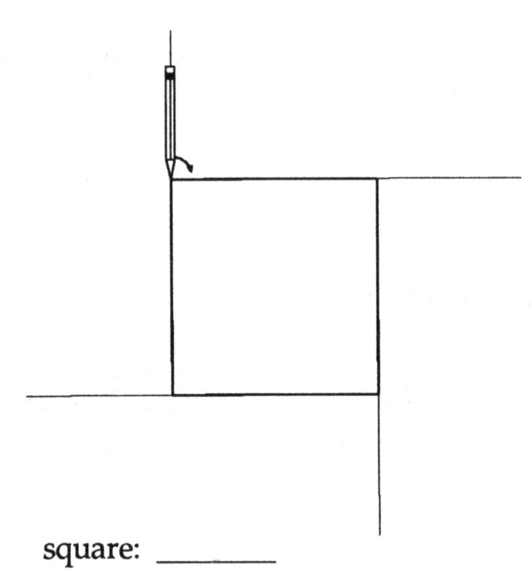

square: _______

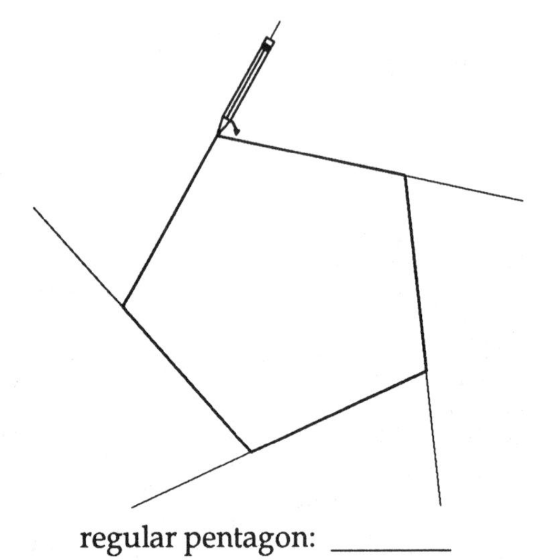

regular pentagon: _______

The conjecture in Exercise 5, when proved, is sometimes known as ***The Exterior Angle Theorem.***

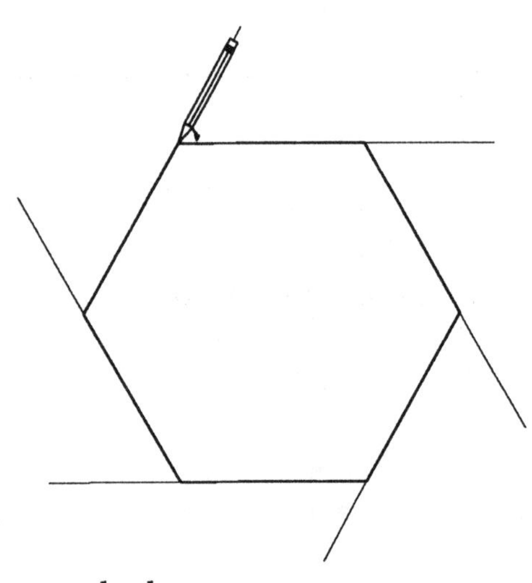

regular hexagon: _______

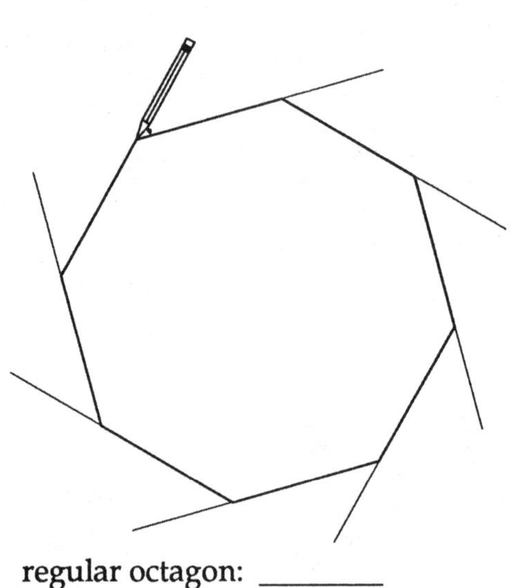

regular octagon: _______

5. State a conjecture based on your findings in Exercise 4:

   The sum of the exterior angles of any regular polygon is ________.

6. The exterior angles of a regular polygon all have the same measure. Use this fact and your conjecture in Exercise 5 to find the measure of each exterior angle for these polygons:

   square: _____ regular pentagon: _____

   regular hexagon: ____ regular octagon: _____

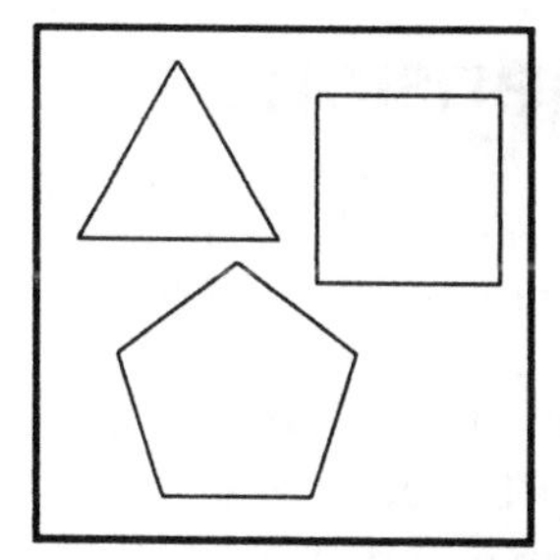

Activity P-1 (continued)

7. Find the measure of each *interior* angle for these polygons:

square: ______ regular pentagon: ______

regular hexagon: _____ regular octagon: ______

The prefix "tri-" means "three," so the word "triangle" refers to a figure with three angles.

Other polygons are named according to the number of angles they have. The suffix "-gon" means "angle" in Greek and actually comes from the Greek word for "knee." The prefixes "penta-" (5), "hexa-" (6), and "octa-" (8), are common in the English language.

In any language, the most commonly used words are often the most irregular. An example of this is the word "square" for a regular quadrilateral.

8. Put your findings in the chart below:

Name of Polygon	**Number of Sides**	**Each Exterior Angle**	**Each Interior Angle**
Equilateral Triangle			
Square			
Regular Pentagon			
Regular Hexagon			
Regular Octagon			

9. Describe a pattern that you see in the chart.

10. Determine the measure of each interior angle of the following regular polygons:

    a. regular nonagon (9 sides)

    b. regular decagon (10 sides)

    c. regular dodecagon (12 sides)

11. Write an explanation for a friend (who missed class) describing how to find the measure of each interior angle of a regular polygon with *n* sides.

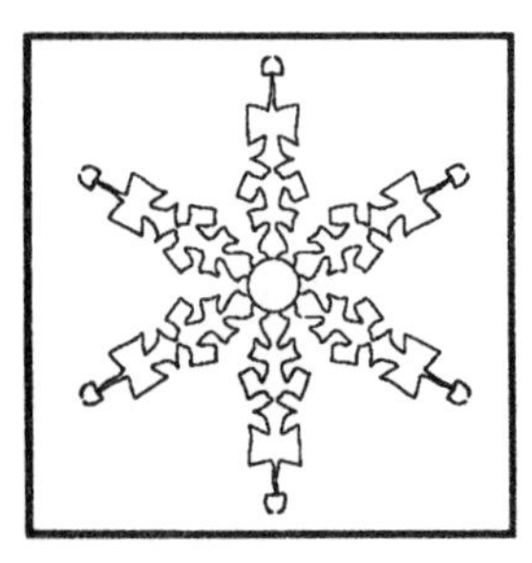

Activity P-2: Symmetry in Two Dimensions

Part I: Rotation Symmetry

The uppercase letter N looks the same right side up or upside down. This letter has **2-fold rotation symmetry**. This means that as you rotate the letter one complete turn (360°) about a point, it appears exactly the same at 2 different positions.

Demonstrate this by placing the tip of your pencil at the center point of the N; now slowly turn the paper halfway around. The N should appear identical to its original self. Make another half turn to return the N to its original position.

N

Rotation symmetry is prevalent in nature. Many microscopic organisms exhibit rotation symmetry. Five-fold symmetry is characteristic of many flowers, while snowflakes provide the best known examples of six-fold symmetry.

Definition: A two-dimensional figure has **rotation symmetry** if it can be turned about a central point (called the **center of rotation**) in such a way that the turned figure appears to be in exactly the same position as the original figure.

Use n to represent the number of times an identical image occurs as the figure is rotated one complete turn (360°). Then we say that the figure has **n-fold rotation symmetry**. (If $n = 1$, we say that the figure has no rotation symmetry.)

1. Consider each of the letters and symbols below:

S A H + × %

a. Which of these figures has no rotation symmetry?

b. Which of these figures have 2-fold rotation symmetry?

Two-fold rotation symmetry is often called half-turn or point symmetry.

c. Which have 4-fold rotation symmetry?

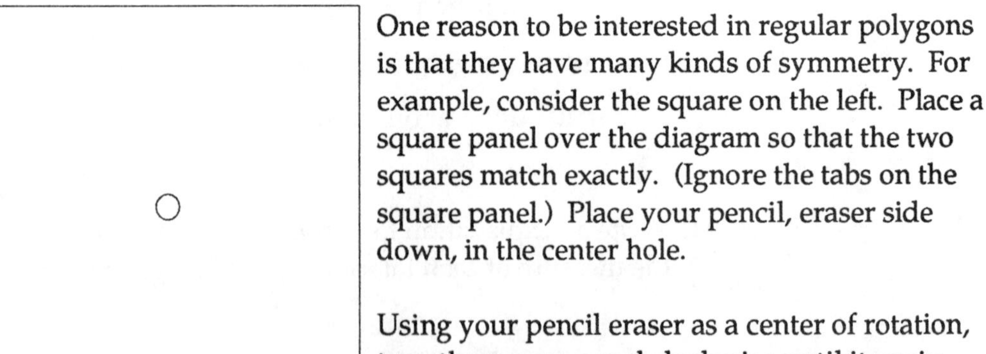

One reason to be interested in regular polygons is that they have many kinds of symmetry. For example, consider the square on the left. Place a square panel over the diagram so that the two squares match exactly. (Ignore the tabs on the square panel.) Place your pencil, eraser side down, in the center hole.

Using your pencil eraser as a center of rotation, turn the square panel clockwise until it again matches up with the diagram.

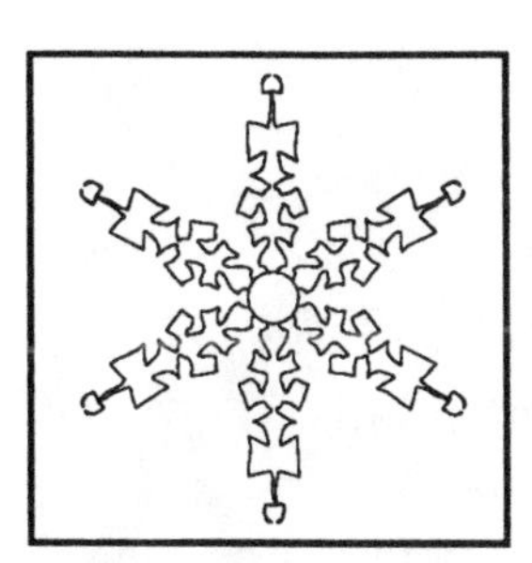

Activity P-2 (continued)

2. a. What fraction of a complete turn did you make?

   b. What is the measure of this rotation in degrees?

   c. A square has ___-fold rotation symmetry.

3. Use the same technique as in Exercise 2 with a pentagon panel.

   A pentagon has ___-fold rotation symmetry.

What kind of rotation symmetry does a circle have?

This property makes it possible to turn a round peg smoothly in a round hole.

4. Make a conjecture: A regular polygon with n sides has ___-fold rotation symmetry about its center.

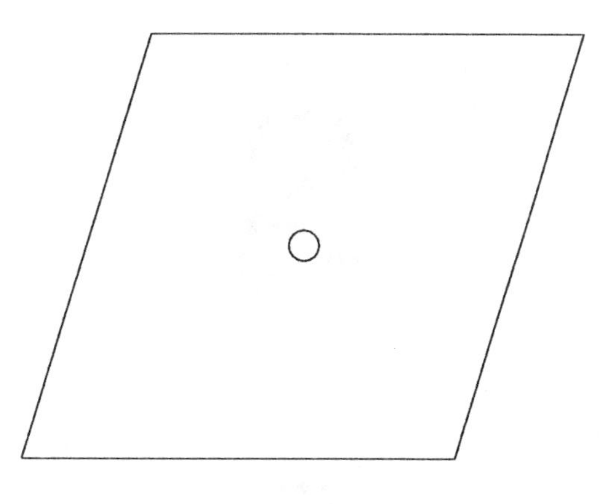

5. Let's try this with a non-regular polygon. Use a rhombus panel.

   a. Why is the rhombus not a regular polygon?

   b. The rhombus has ___-fold rotation symmetry.

6. Does your conjecture in Exercise 4 hold for non-regular polygons with n sides?

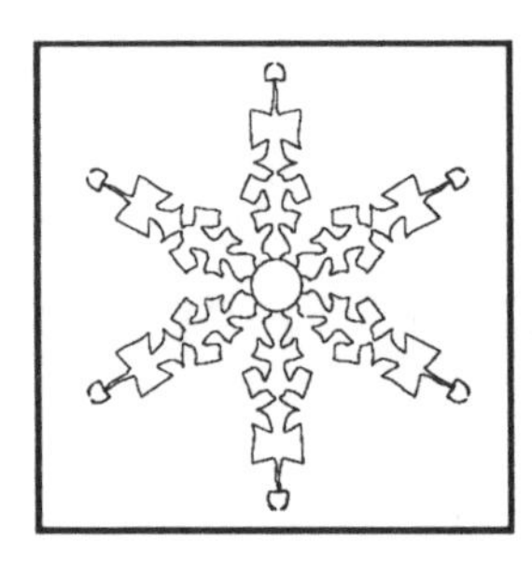

Activity P-2 (continued)

Part II: Reflection Symmetry

The uppercase letter A has **reflection symmetry** about the dotted line shown. To see this, fold your paper along the dotted line. Note that the two halves of the folded A match exactly. The fold in your paper demonstrates an **axis of reflection** for the A.

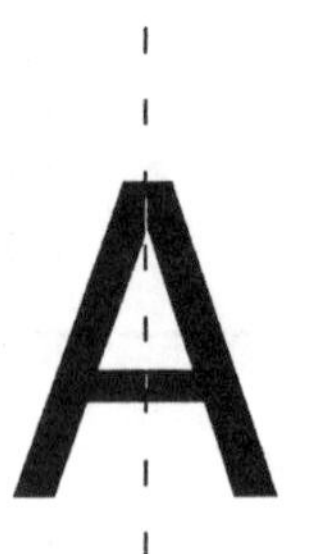

*Reflection symmetry is sometimes called **line symmetry**, or **bilateral symmetry**. Most animals, including people, exhibit bilateral symmetry, at least externally. Notice that the right side of your body is pretty much a mirror image of the left side.*

Definition: A two-dimensional figure has **reflection symmetry** about a line (called the **axis of reflection**) if the pattern on one side of the line is the mirror image of the pattern on the other side.

7. Find two axes of reflection for the H in the diagram. Sketch these lines on the picture.

*The axis of reflection is sometimes called a **mirror line**.*

8. Sketch all axes of reflection, if any, for each of the letters and symbols below.

E S X

/ T O

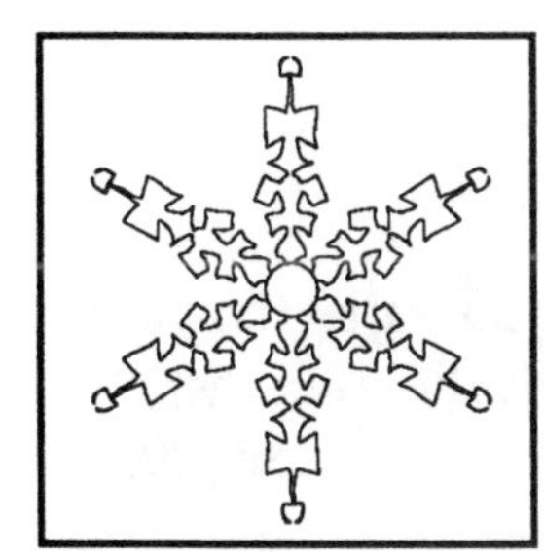

Activity P-2 (continued)

Regular polygons have several axes of reflection. In the following exercise you will make a square, complete with its axes of reflection.

9. Take any rectangular piece of paper and, by matching opposite edges, fold it carefully into quarters. Hold it so that the center of the original paper is at the bottom right. Now fold it one more time so that you bisect the 90° angle at the bottom right. Cut off the excess paper so edges match.

   Now open up your paper. You have a square in which each fold lies on an axis of reflection.

   a. How many axes of reflection does your square have?

   b. Is this true for any square?

A mirror kaleidoscope can be used to form the image of a regular polygon from just one line segment.

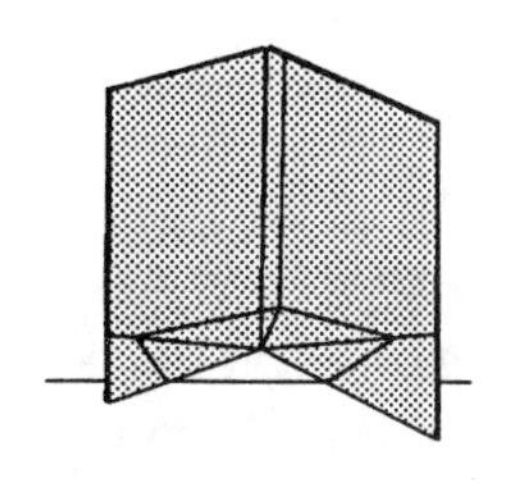

Other regular polygons can be seen by changing the angle between the mirrors.

10. Find as many axes of reflection as you can for a triangle, pentagon, and rhombus panel. Draw them with ballpoint pen on your polygon panels and fold along the lines to see if you are right. Then sketch the reflection lines on the shapes below.

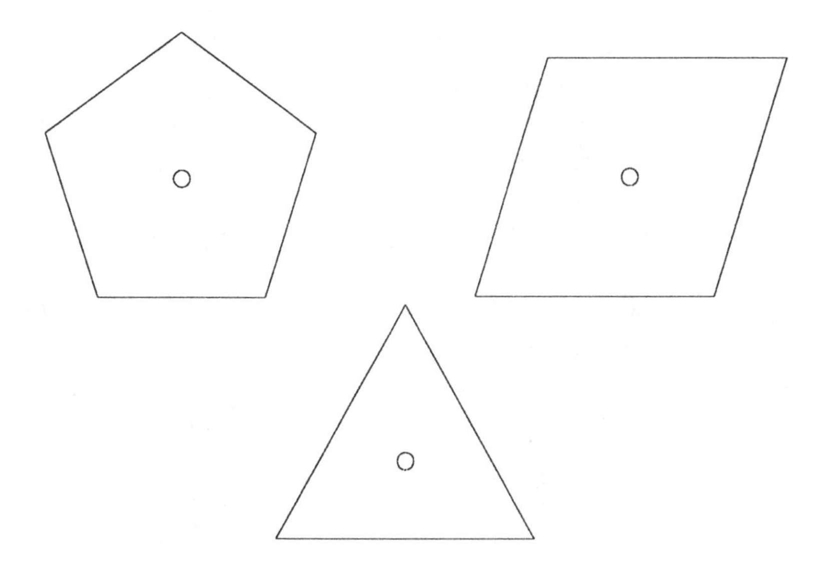

11. Make a conjecture: A regular polygon with n sides has ___ reflection axes.

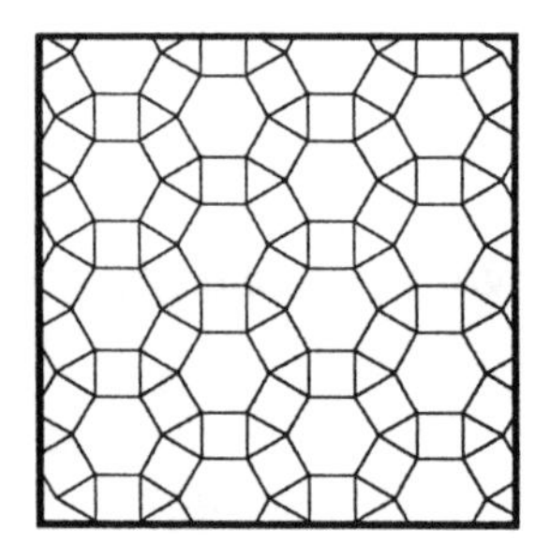

Activity P-3: Polygon Tessellations

Arrange several of the square panels as if tiling a floor. Make sure that the panels do not overlap and that the covering has no gaps, except for the holes in the centers and at the corners of the panels. Notice that this pattern of square panels could be extended indefinitely.

Examples of tilings can be found in floor patterns and on wallpaper. Beautiful mosaic tilings can be seen in Moorish mosques and palaces, such as the Alhambra in Granada, Spain.

Definition: An arrangement of closed shapes which covers the plane without gaps or overlaps is called a **tessellation** or **tiling**.

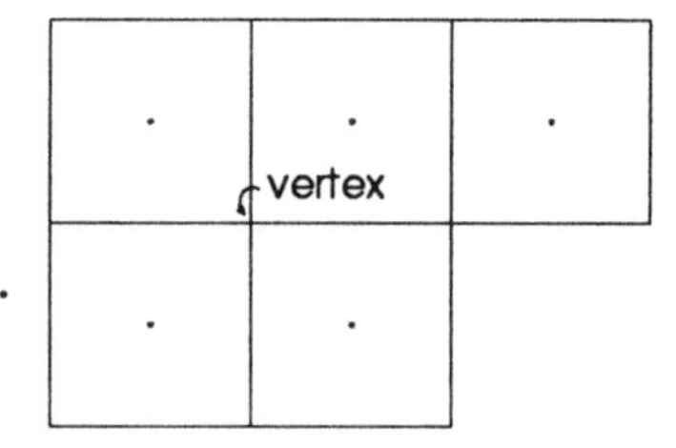

1. Put your finger on one vertex in your tessellation which is completely surrounded by squares.

   a. How many squares surround your finger?

   b. How many angles share that vertex?

   c. What is the total angle measure around that vertex?

There are several ways to form tessellations using polygons. The one you have made with square panels is an example of a **regular tessellation**, a tessellation using only one regular polygon shape.

The Dutch artist M.C. Escher (1898-1972) was fascinated by tilings. His brilliant idea was to form repeated patterns of recognizable figures, such as birds, fish, or reptiles. Below is an Escher-like tiling.

2. Experiment with other regular polygon panels. What other regular tessellations can you make?

3. Explain why you cannot make a regular tessellation using only regular octagons.

4. Can you make a tessellation using only the rhombus panels?

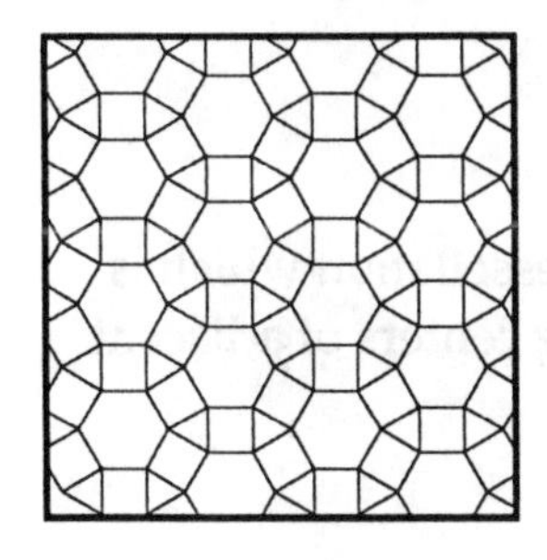

Activity P-3 (continued)

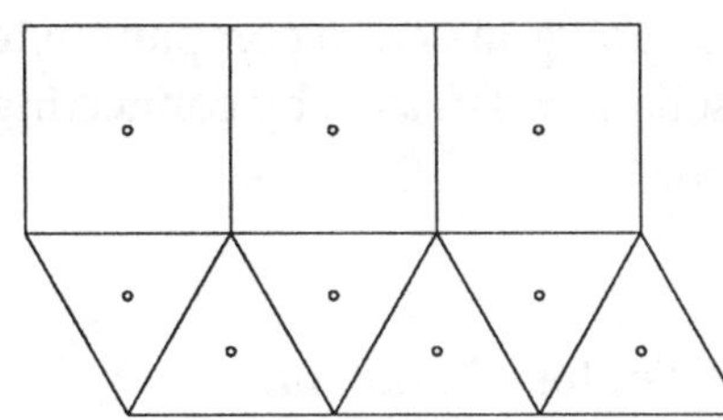

A tessellation that uses more than one type of regular polygon as its tiles is called a **semiregular tessellation** if the arrangement of tiles at each vertex is the same. Use triangles and squares to make the semiregular tessellation shown in the diagram.

Regular pentagons don't tile, but many equilateral (though not equiangular) pentagons do.

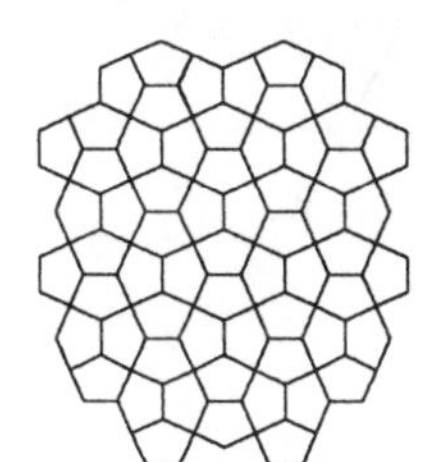

This pattern is seen in street tiling in Cairo and in the mosaics of Moorish buildings. A similar tiling can be obtained as the dual of a semi-regular tiling (see exercise 8).

5. Put your finger on one vertex in your tessellation of squares and triangles.

   a. Describe the arranglement of the shapes surrounding your finger.

   b. How many angles surround that vertex?

   c. What is the total angle measure around that vertex?

   d. Choose another vertex in the tessellation. Is the arrangement of the tiles surrounding that vertex the same as in 5a?

6. a. Find a different arrangement of triangles and squares that forms a semiregular tessellation. Sketch part of this tessellation. Describe the order of the shapes that surround each vertex.

   b. Experiment with the regular polygon panels to find other examples of semiregular tessellations. Make a sketch on the back of this sheet of each one that you find.

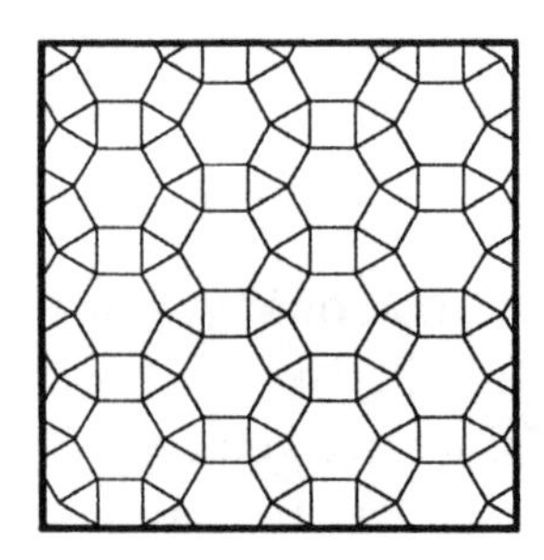

Activity P-3 (continued)

Each tessellation using regular polygons has a companion tessellation which is called its **dual**. The dual tessellation is drawn by connecting centers of adjacent tiles in the original tessellation.

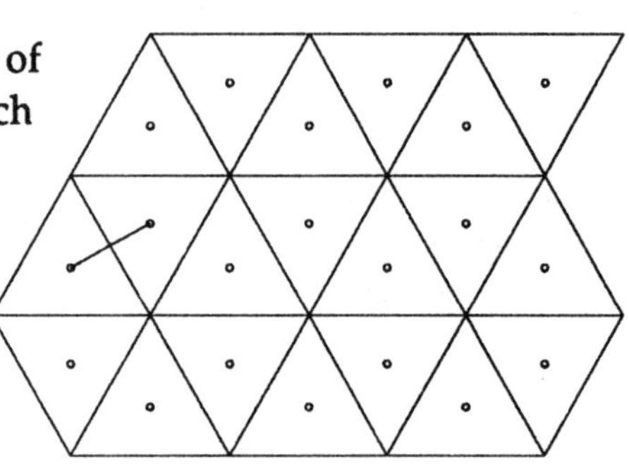

7. This diagram shows a regular tessellation of equilateral triangles, with the center of each triangle marked. Use a straightedge and pencil to connect the centers of two adjacent triangles. Continue to do this for the entire tessellation. The new tessellation you have made is the **dual** of the original tessellation.

   What shapes are the tiles in the dual tessellation?

When drawing the dual of a tessellation, the line segment connecting the centers of two adjacent shapes must cross the common edge of the two shapes.

8. Draw the dual of the semiregular tessellation shown by connecting the centers of adjacent polygons.

The dual of a regular tessellation will be a regular tessellation. What about the dual of a semiregular tessellation?

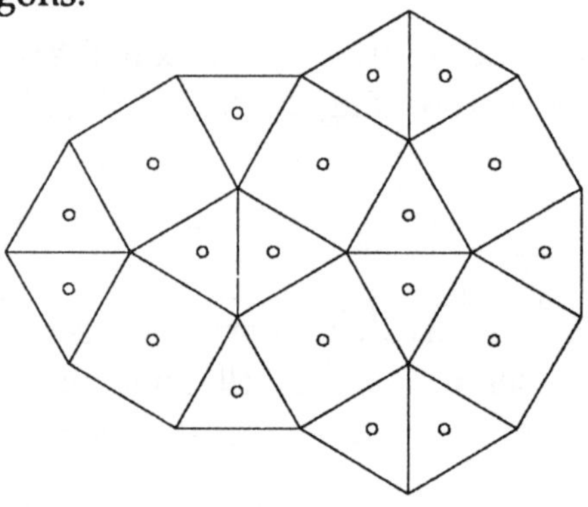

   a. Is the dual tessellation a regular tessellation?

   b. Is it a semiregular tessellation?

   c. What shapes are the tiles in the dual tessellation?

   d. How many tiles are at each vertex of the original semiregular tessellation?

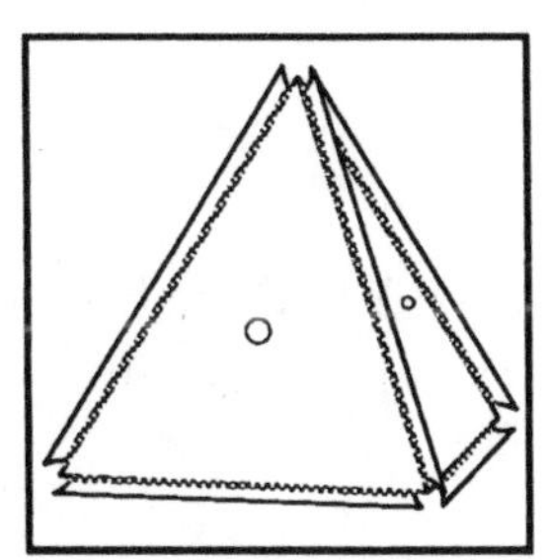

Activity 1: The Regular Polyhedra

*A polyhedron is a three-dimensional shape formed by joining edges of polygons to enclose a region of space. The polygons are called the **faces** of the polyhedron. Exactly two polygons meet at each **edge** of the polyhedron. At least three faces meet at each **vertex** of the polyhedron.*

1. Using some of the polygon panels, make a polyhedron in the following way:
   - Connect the edges of two polygons by holding the tabs together with one hand and fastening a rubber band over them with the other hand.
   - Continue to connect polygons together until you have completely enclosed some region of space.
   - You must be sure that there are at least three faces at each vertex and exactly two faces at each edge of the polyhedron.

Describe your polyhedron:

a. How many faces are there?

b. Which polygons did you use for the faces?

c. What configurations of polygons do you have at the vertices?

*Two polygons are **congruent** if their corresponding sides are equal in length and their corresponding angles are equal in measure. Two congruent polygons have the same size and shape.*

Definition: A **regular polyhedron**, or Platonic solid, is a polyhedron with the following properties:

a. All faces are regular polygons.

b. All faces are congruent to each other.

c. The same number of faces meet at each vertex in exactly the same way.

2. Examine the polyhedron you made in Exercise 1. Is it a regular polyhedron? If it is not a regular polyhedron, explain which of the necessary properties is (are) missing.

3. Build as many different regular polyhedra as you can using squares only. Remember that you must have the same number of faces at each vertex. How many can you find?

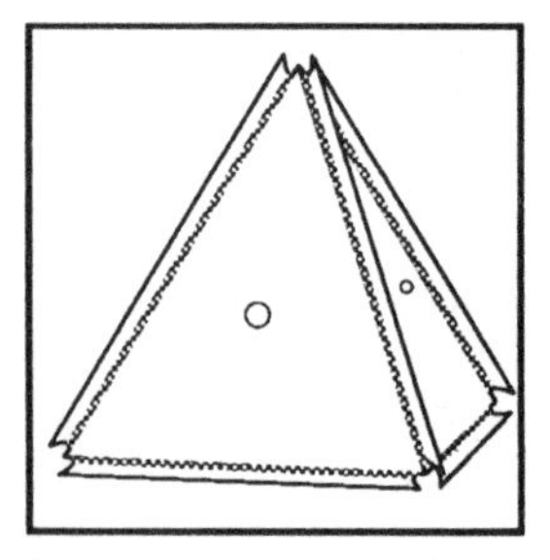

Activity 1 (continued)

4. Build as many different regular polyhedra as you can using only equilateral triangles for faces. How many can you find?

5. Build as many different regular polyhedra as you can using only regular pentagons for faces. How many can you find?

Dice made in the shape of regular polyhedra are used in games such as Dungeons and Dragons™ because each face has an equal likelihood of landing flat. The cube is the most common shape for dice. In fact, the word for dice in Greek is 'cubos.'

6. Why is it impossible for a regular polyhedron to have faces which are regular hexagons? (Remember that at least three faces must meet at each vertex.)

7. Why is it impossible for a regular polyhedron to have faces with more than six sides?

8. Your answers in Exercises 3-7 should convince you that there are exactly five regular polyhedra. Their names describe how many faces they have.

   a. For each polyhedron, count the number of faces it has and enter this number in the first column in the chart below.

Number of Faces	Name	Kind of Faces	Number of Faces at Each Vertex

Here are some other polyhedra names that are based on the number of faces of the polyhedron:

pentahedron—5 faces
decahedron—10 faces
tetrakaidecahedron—14
pentakaidecahedron—15
hexakaidecahedron—16

What do you suppose the word "kai" means?

   b. From the following list, choose the appropriate name for each shape. The prefix of each name tells the number of faces a polyhedron has. (You may want to use a dictionary to help you.) Enter the names in the chart.

   Names: dodecahedron, tetrahedron, hexahedron, octahedron, icosahedron

   c. To complete the chart, note the kind of face (triangle, for example) that the polyhedron has and the number of faces that meet at each vertex.

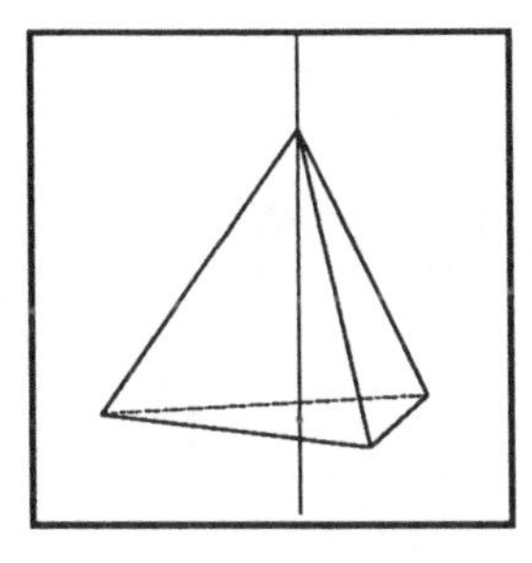

Activity 2: Symmetry of Regular Polyhedra

The five regular polyhedra are known as Platonic solids, after the Greek philosopher Plato. Plato was so taken with the beauty and regularity of the solids that he used them to represent the elements thought to make up the world.

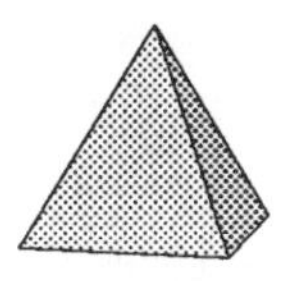 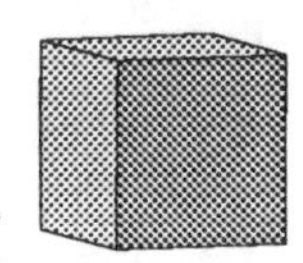 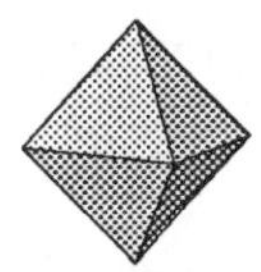 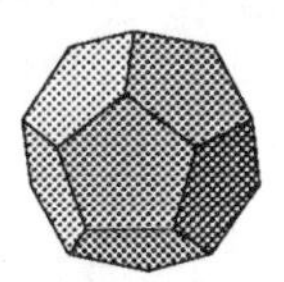 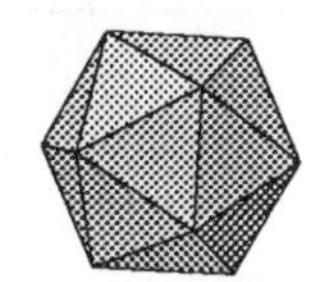

Plato (b. 429 B.C.) was a Greek philosopher. Even though the regular polyhedra were known well before his time, they are referred to as ***Platonic solids*** *because he wrote about them in his dialogue* Timaeus. *In this allegory, the polyhedra represent the elements of the physical world.*

1. Look at your models of the five Platonic solids. What would you say is "regular" (uniform) about each one?

One property that makes the shapes of the Platonic solids so pleasing is their symmetry. In the exercises below, you will investigate the symmetries of some of the Platonic solids.

Rotation Symmetry

Examples of Platonic solids abound in nature. Crystals of pyrite are often shaped like cubes; alum crystals occur as octahedra. The atoms of a methane molecule are arranged like the corners of a tetrahedron.

Many viruses are shaped like icosahedra, while the skeletons of some miscroscopic animals called radiolaria *are shaped like dodecahedra, octahedra, or icosahedra.*

A three-dimensional figure is turned about an **axis of rotation** to determine its rotation symmetry.

2. Pick up your regular octahedron and put a pipe cleaner or straw through a pair of opposite vertices. Hold the straw vertically and look directly down on it; now turn the octahedron a quarter turn. You should see that the octahedron looks as if you had not turned it—faces, vertices, and edges appear to be in the same position as before. The straw demonstrates an axis of **4-fold rotation symmetry**.

   How many different axes of 4-fold rotation symmetry does the octahedron have? (Hint: How many pairs of vertices are there?)

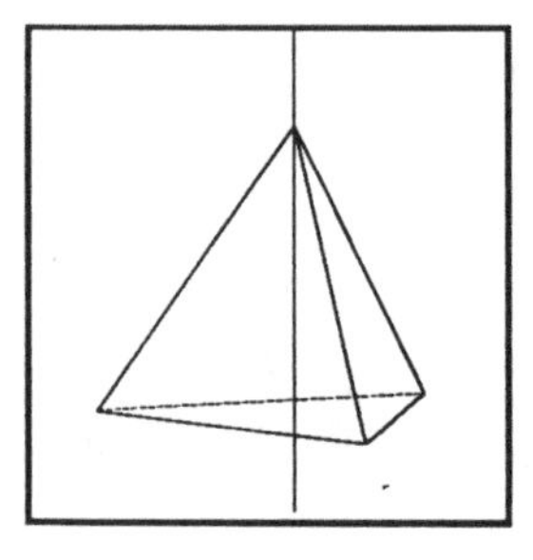

Activity 2 (continued)

3. The cube also has axes of 4-fold rotation symmetry, but they do not go through the vertices. Holding your cube, put a pipe cleaner or straw through the centers of two opposite faces. Turn the cube around this axis 1/4 of a turn; it looks just like it did in its original position.

   How many axes of 4-fold rotation symmetry does the cube have?

4. The cube has axes of 3-fold rotation symmetry as well. Use a pipe cleaner or straw to connect diagonally opposite corners of the cube, piercing through the center of the cube. Hold the axis (the straw) vertically and look directly down on the model. Slowly turn the axis until the faces, vertices, and edges of the cube first appear to be in the same position as before.

   a. How far have you turned?

   b. How many axes of 3-fold rotation symmetry does the cube have?

Why are the symmetries of the cube and octahedron alike? It is because they are ***dual polyhedra****; the vertices of one correspond to the faces of the other, and vice versa. The dual relationship of a pair of objects often serves as a shortcut, allowing us to determine the properties of both at the same time.*

5. Use a pipe cleaner or straw to demonstrate an axis of 3-fold rotation symmetry on your regular octahedron.

   a. Where is it located?

   b. How many such axes of 3-fold rotation symmetry does the octahedron have?

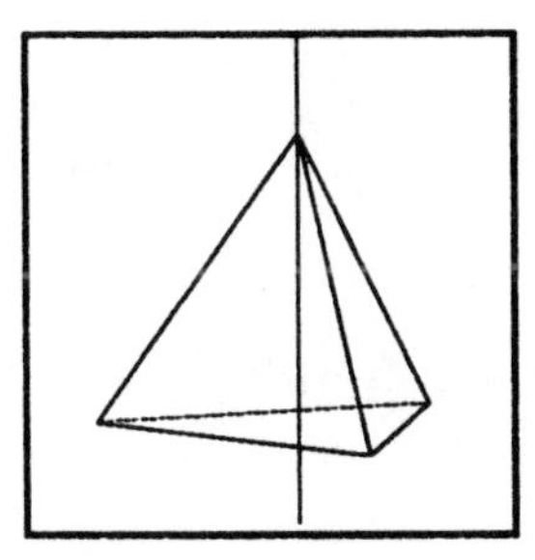

Activity 2 (continued)

6. Use a pipe cleaner or straw to demonstrate an axis of 3-fold rotation symmetry on your regular tetrahedron.

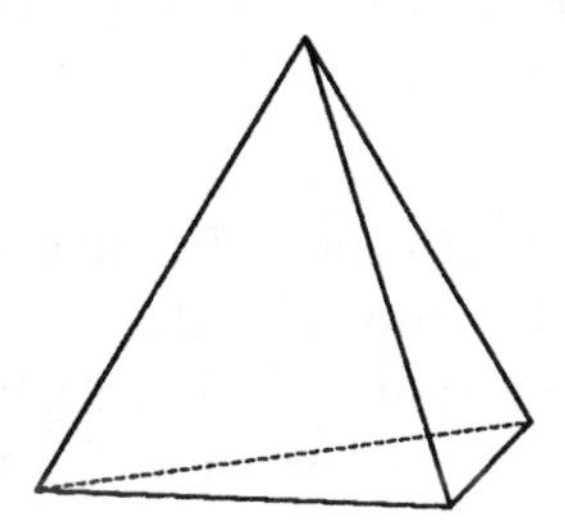

   a. Where is it located? Draw it on the picture shown here.

   b. How many such axes of 3-fold rotation symmetry does the tetrahedron have?

7. The cube, the octahedron, and the tetrahedron have another kind of rotation symmetry, in which the axis of rotation connects the midpoints of two opposite edges and passes through the center of the figure. Use a straw to demonstrate one such axis of rotation on each polyhedron.

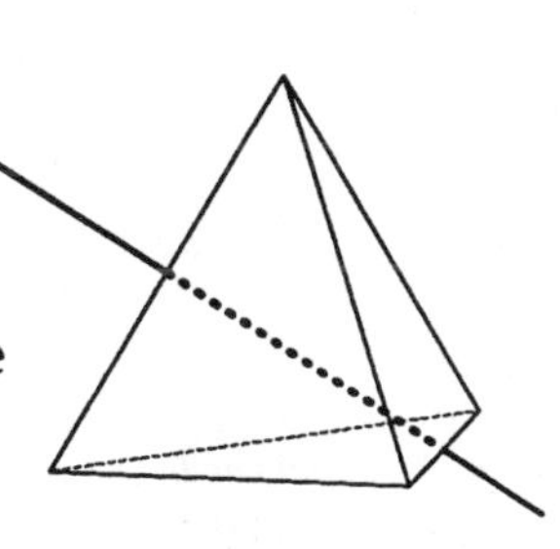

The tetrahedron does not have opposite vertices or opposite faces, so a regular tetrahedron has only two kinds of rotation symmetry.

   a. Hold the straw vertically and look directly down on the model as you turn it. What kind (?-fold) of rotation symmetry does this axis demonstrate on the tetrahedron?

      Cube?

      Octahedron?

   b. How many such axes of symmetry does the cube have?

   c. How many such axes of symmetry does the octahedron have?

   d. How many such axes of symmetry does the tetrahedron have?

8. Summarize you findings about **rotation symmetry** by entering your answers to Exercises 2-7 in the following table.

Name	# of 4-fold Axes	# of 3-fold Axes	# of 2-fold Axes
Cube			
Octahedron			
Tetrahedron			

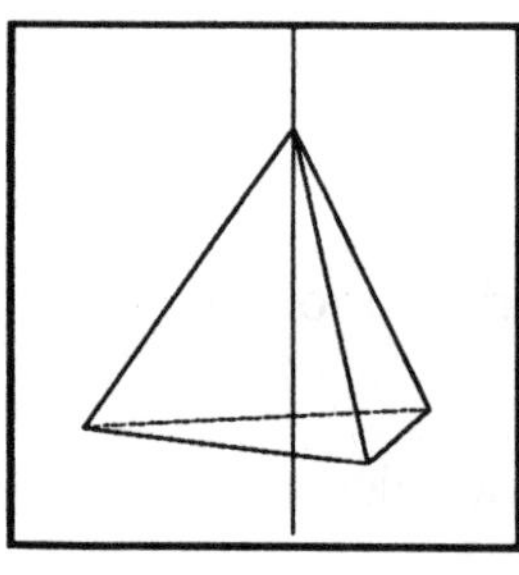

Activity 2 (continued)

Reflection Symmetry

A three-dimensional figure has **reflection symmetry** about a plane if the plane cuts the figure into two mirror-image shapes.

Each Platonic solid has several planes of reflection. In the exercises below, you will investigate the reflection planes of the cube and the regular octahedron. Build the transparent cube from an acetate net (net page 2), taping the edges with transparent tape. Use the regular octahedron from Activity 1, and colored rubber bands.

9. Demonstrate one plane of reflection of the cube by wrapping a rubber band *across* four parallel edges, as shown. The rubber band must bisect each face it crosses into two mirror-image halves.

The rubber band shows the intersection of the cube with one of its reflection planes.

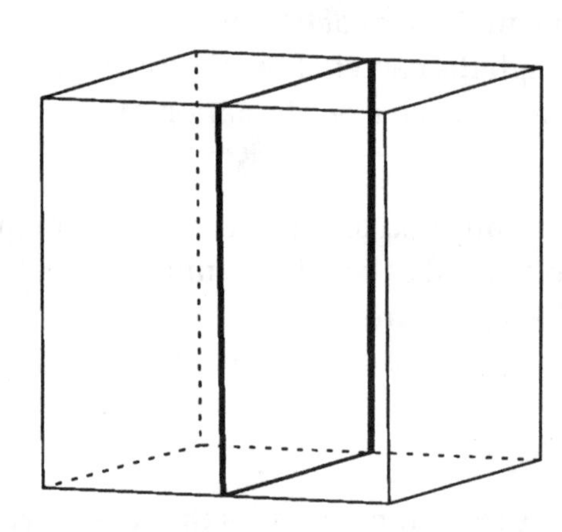

a. How many edges does the cube have?

b. How many of these reflection planes does the cube have?

c. Put additional rubber bands on your cube to show all these reflection planes, and sketch them on the picture.

10. The cube has another set of reflection planes. Demonstrate one such plane by wrapping a colored rubber band along two opposite edges and across two faces, as shown.

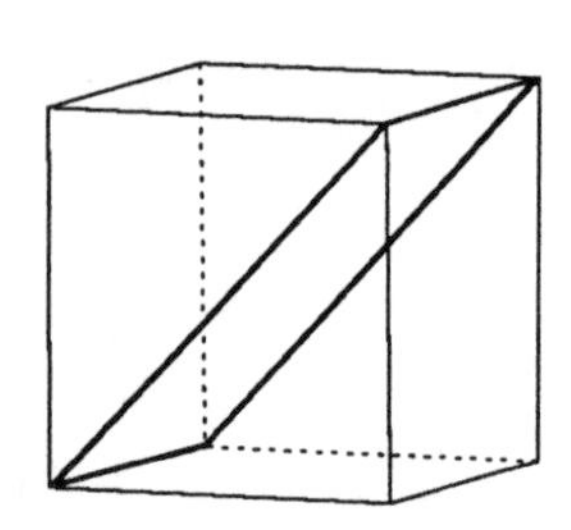

How many of these planes of reflection symmetry are there?

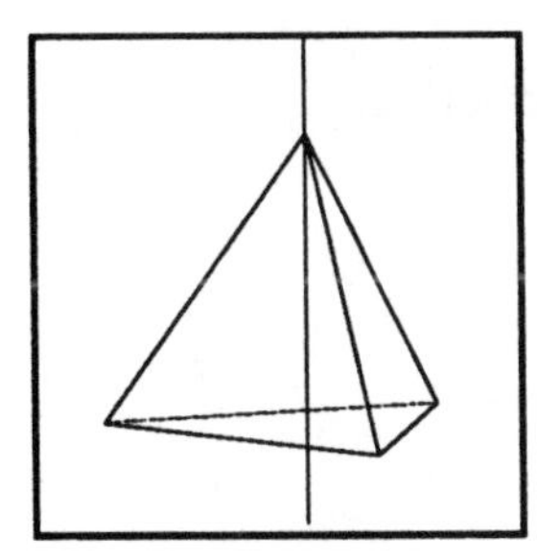

Activity 2 (continued)

11. Take all the rubber bands off, open the cube and place the octahedron inside so that each vertex of the octahedron touches the center of a face of the cube. Each plane of reflection for the cube will also be a plane of reflection for the octahedron.

    a. Place one rubber band on the cube as in Exercise 9. Sketch this reflection plane on the octahedron below.

    b. Place another rubber band on the cube as in Exercise 10. Sketch this reflection plane on the octahedron.

One kind of reflection plane of the octahedron cuts the shape into two pyramids. The pyramids of ancient Egypt were not exactly half-octahedra, however, as the slope of their walls was slightly less steep than the walls of a regular octahedron.

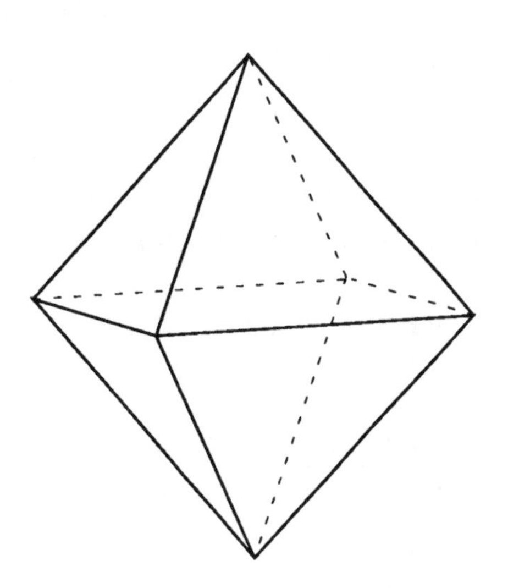

Activity 3: Other Symmetric Polyhedra

A polyhedron with the following properties is **regular**:

a. All faces are regular polygons.

b. All faces are congruent to each other.

c. The same number of faces meet at each vertex in the same way.

The most familiar prism shape is a triangular prism. A glass triangular prism separates light into a rainbow spectrum of colors.

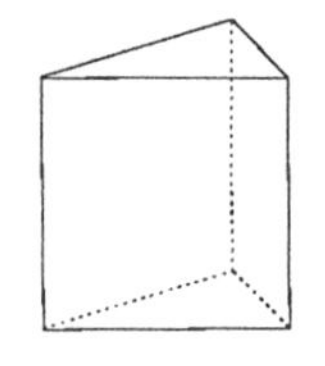

If a polyhedron lacks any of these properties, then it is not regular. In this activity you will build several polyhedra that are not regular.

1. Use one square panel and four triangle panels to construct a square pyramid, as shown in the picture.

   This polyhedron is not regular. Which properties above does it lack?

An antiprism has two congruent regular polygonal bases joined by equilateral triangles.

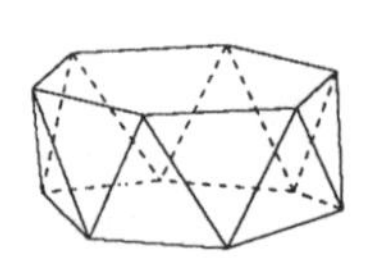

Drums are often laced like antiprisms. The top polygon is rotated with respect to the bottom polygon, giving the antiprism a twist.

2. A **regular right prism** is formed by two congruent regular polygonal bases which lie in parallel planes and have corresponding edges joined by squares that are perpendicular to the bases.

   Use two triangle panels as bases and connect them with three square panels to construct a regular right triangular prism.

   Use two pentagon panels as bases and connect them with five square panels to construct a right pentagonal prism.

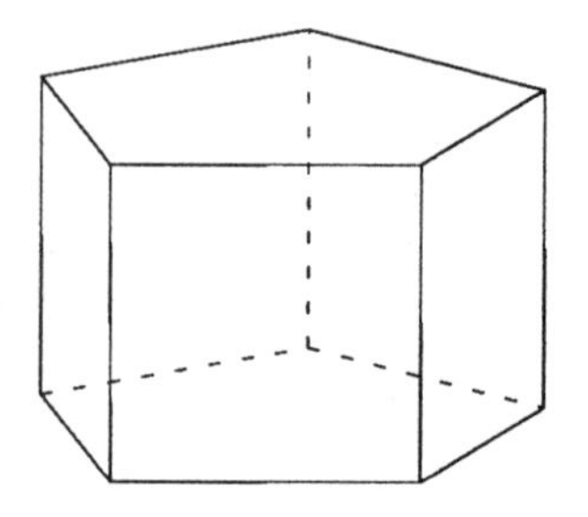

   a. The two prisms you have made are not regular polyhedra. Which property for regular polyhedra do they lack?

   b. One of the Platonic solids is also a right prism, but it is not usually called a prism. Which one is it?

One of the Platonic solids is also an antiprism. Which one?

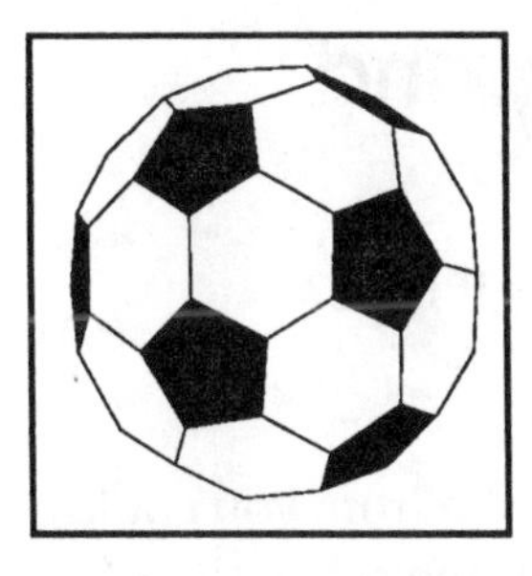

Activity 3 (continued)

If a polyhedron has properties a and c above but lacks property b, then it is called **semiregular**. All prisms and antiprisms are semiregular polyhedra. There are 13 other semiregular solids; these are known as the **Archimedean solids**. Like the Platonic solids, all vertices of an Archimedean solid are exactly the same, and the faces are regular polygons, but there are two or more different types of faces.

The Archimedean solids are named after Archimedes, a Greek who lived in the 3rd century B.C. He is the mathematician who, according to legend, jumped naked from the bathtub and ran through the streets shouting "Eureka!" when he discovered that a body placed in water will displace an amount of water equal to the weight of the body. Another legend says that he was killed by Roman soldiers when he refused to accompany them until he had finished the geometry problem on which he was working.

3. One Archimedean solid is the **cuboctahedron**. Build one by connecting two squares and two triangles, alternately, at each vertex. Continue until you have a closed shape.

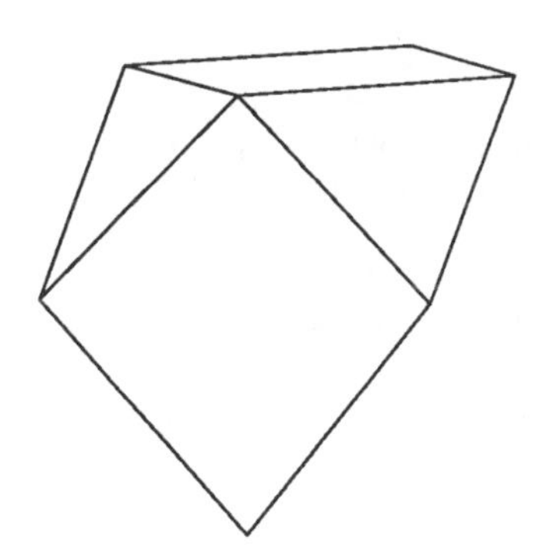

   a. How many squares did you use altogether?

   b. How many triangles did you use altogether?

   c. Explain why the shape has the name cuboctahedron.

The cuboctahedron is an Archimedean solid. Another one is the truncated icosahedron, which is pictured at the top of the page.

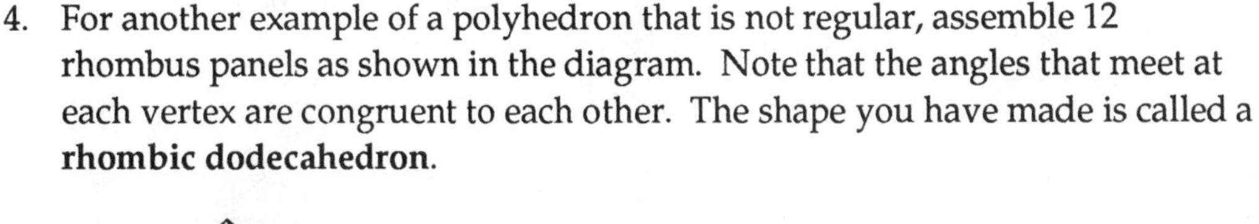

4. For another example of a polyhedron that is not regular, assemble 12 rhombus panels as shown in the diagram. Note that the angles that meet at each vertex are congruent to each other. The shape you have made is called a **rhombic dodecahedron**.

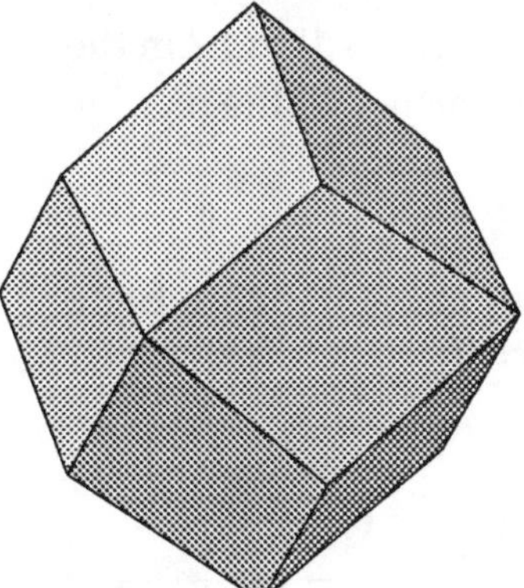

   a. Which properties of a regular polyhedron does this shape lack?

   b. Is it a semiregular polyhedron? Explain.

   c. Explain why the shape has the name *rhombic dodecahedron*.

If you pack twelve congruent spheres around a single congruent sphere, as if you were making a pile of snowballs, the centers of the outside spheres will form a cuboctahedron.

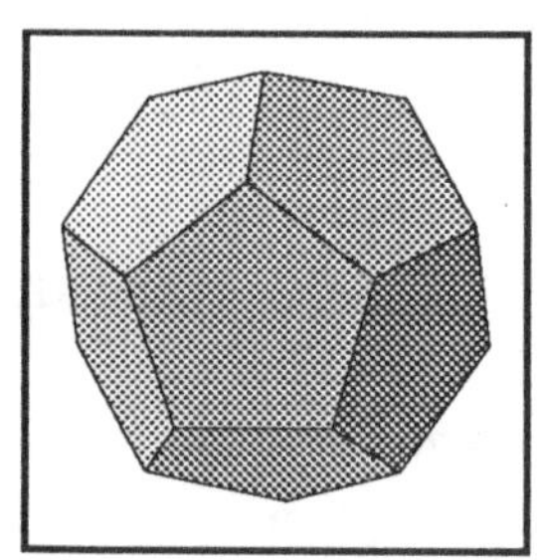

Activity 4: Counting Faces, Edges, and Vertices of Polyhedra

1. Look at the regular dodecahedron. How many faces does it have?

In order to find the number of *edges* on the regular dodecahedron, you could just count them. But the regularity of this polyhedron allows us to use a shortcut.

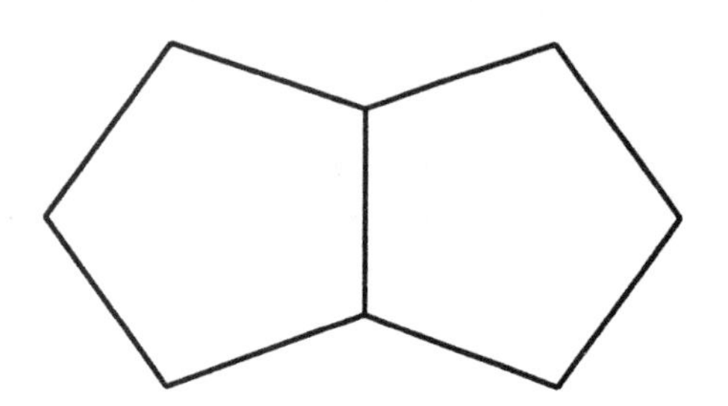

Note that there are 5 edges on each face, and that exactly 2 faces share an edge. Therefore, there are (12 x 5)/2 = 30 edges on the regular dodecahedron.

The angle between two faces of a polyhedron is called a ***dihedral angle****. All dihedral angles of a regular polyhedron have the same measure.*

When you count each edge on each face of a polyhedron, you are counting each edge of the polyhedron twice.

When you count each vertex on each face of a polyhedron, you are counting each vertex of the polyhedron more than once, depending on the number of faces which meet at each vertex.

2. You can use a similar shortcut to find the number of *vertices* on the regular dodecahedron.

a. How many vertices does each face have?

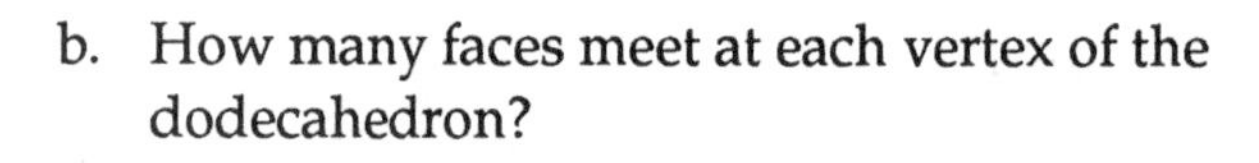

b. How many faces meet at each vertex of the dodecahedron?

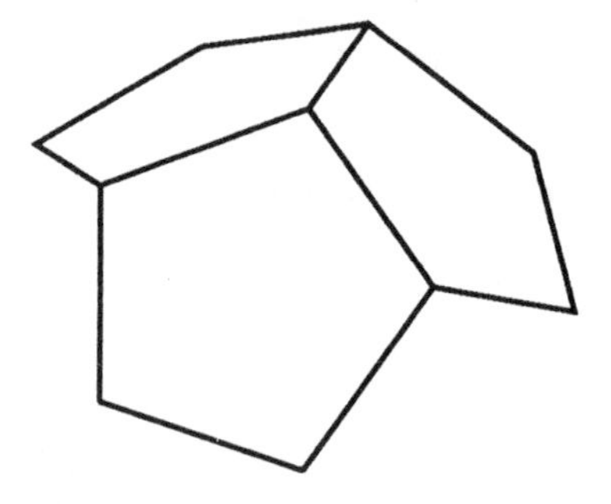

c. How many vertices are there on the regular dodecahedron?

3. Enter the information from Exercises 1 and 2 in the chart below. Fill in the chart in a similar manner for the rest of the regular polyhedra that you have built.

Name	Number of Faces (f)	Number of Vertices (v)	Number of Edges (e)
Tetrahedron			
Hexahedron (cube)			
Octahedron			
Dodecahedron			
Icosahedron			

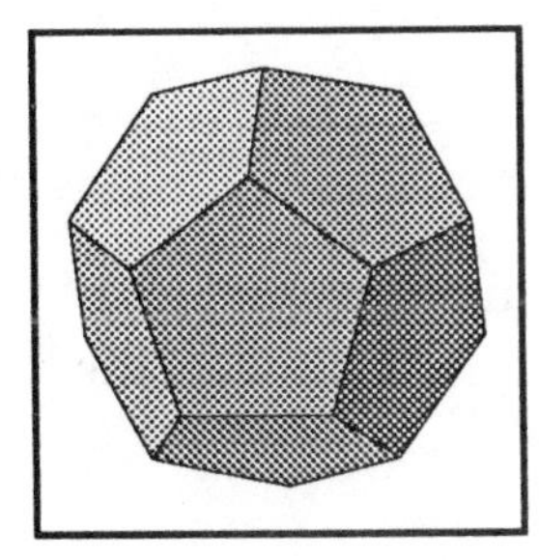

Activity 4 (continued)

4. Fill in the chart below for the polyhedra listed. When possible, find shortcuts to count edges and vertices.

Name	Number of Faces (f)	Number of Vertices (v)	Number of Edges (e)
Square Pyramid			
Right Triangular Prism			
Right Pentagonal Prism			
Cuboctahedron			
Rhombic Dodecahedron			

Numbers alone do not determine the shapes of polyhedra. For example, a pentagonal prism has exactly the same number of faces, vertices, and edges as the shape which is obtained by cutting off one corner of a cube.

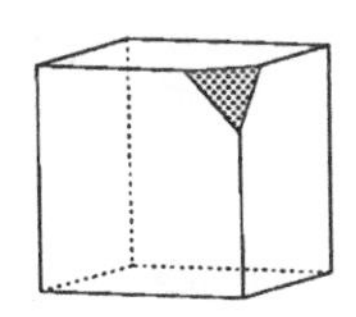

However, this shape differs considerably from the pentagonal prism in the shape of the faces and the way in which the faces are connected.

5. There are several patterns in the charts in Exercises 3 and 4. Describe a pattern that you see.

Activity 5: Euler's Formula and Other Relationships

1. For each regular polyhedron, add the number of faces (f) and the number of vertices (v), and compare this sum to the number of edges (e). Complete the following equation to describe the relationship:

$$f + v =$$

Leonard Euler (pronounced "Oi•ler"), an 18th century Swiss mathematician, was one of the most productive mathematicians of all time. Euler loved children and often worked with a baby on his lap and some of his other children playing around him (he had 13, though 8 of them died while still young).

This relationship is known as **Euler's Formula**.

2. Write down the names of some other polyhedra which fit Euler's Formula. (Check the ones in the second chart in Activity 4. Perhaps you can find others, as well.)

3. Euler's Formula applies to any polyhedron which, if made of rubber, could be blown up into a sphere. Euler's Formula does *not* apply to polyhedra which have tunnels or holes in them. Demonstrate this by finding the numbers of faces, vertices, and edges for the polyhedron drawn below. (Its faces are all trapezoids or rectangles.)

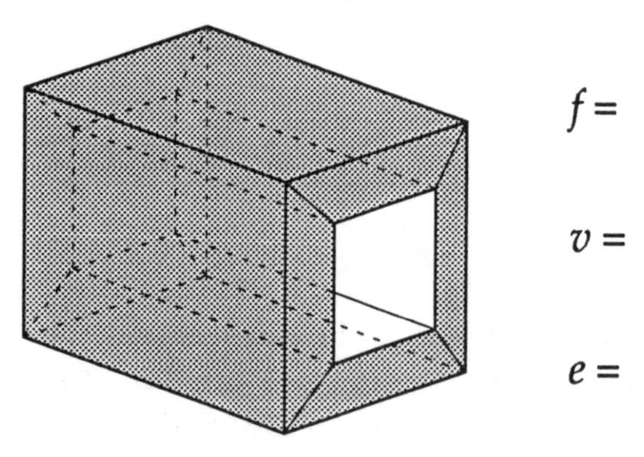

$f =$

$v =$

$e =$

A hat, called a mazzocchio, had the shape of a polyhedron doughnut and was worn by men in the time of the Renaissance.

4. The picture on the right shows the microscopic sea creature called *aulonia hexagona*. It is often described as having hexagonal cells with three hexagons meeting at each vertex. Look carefully at the picture, and you'll see that some of the cells are not hexagons. What other shapes do you see?

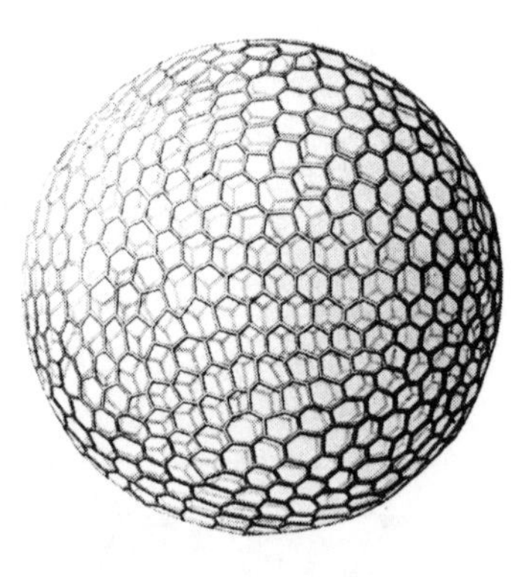

Activity 5 (continued)

5. Euler's Formula can be used to prove that there is no convex polyhedron with just hexagonal faces, three meeting at each vertex.

   For the proof, we suppose there is such a polyhedron and then show that it's impossible.

Remember that when you multiply the number of edges on each face by the number of faces, you have counted each edge of the polyhedron twice.

   a. Suppose there is a convex polyhedron with only hexagonal faces, three meeting at each vertex. Let f be the number of faces on this polyhedron. Find an expression for the number of edges (e) in terms of f.

   b. Find an expression for the number of vertices (v) in terms of the number of faces (f).

   c. If the polyhedron is convex, Euler's Formula must be true for the faces, vertices, and edges of the polyhedron. Use the expressions obtained in a and b to substitute for v and e in Euler's Formula. Is the equation satisfied?

   When a polyhedron does not satisfy Euler's Formula, it cannot be convex.

6. Even when you cannot directly count faces, edges, and vertices of a polyhedron, the properties of the polyhedron determine certain relationships between these numbers. Here's one such relationship which is true for *any* polyhedron with f faces, v vertices, and e edges: $2e \geq 3f$.

   Here's how to obtain this inequality:

   a. Each face of the polyhedron must have three or more edges. Why?

Activity 5 (continued)

b. Add up all the edges of all the faces; the sum of all these edges is greater than or equal to $3f$. But this sum also equals twice the number of edges on the polyhedron. Why?

Therefore, $2e \geq 3f$.

c. Check that this inequality holds for the polyhedra that you have built. For which of the Platonic solids is $2e$ exactly equal to $3f$?

The inequalities $2e \geq 3f$ and $2e \geq 3v$, together with Euler's Formula, can be used to show that every convex polyhedron must have at least 4 faces, 4 vertices, and 6 edges. Thus, the tetrahedron is the "minimum" polyhedron.

Similarly, no convex polyhedron with 7 edges can exist. The three relationships here can be used to prove this.

7. A similar inequality gives a relationship between the number of edges e and the number of vertices v of any polyhedron.

a. In any polyhedron, at least three edges meet at each vertex. Why?

b. Add up all the edges that meet at each of the vertices; the sum of all these edges is greater than or equal to $3v$. But this sum also equals twice the number of edges on the polyhedron. Why?

Therefore, $2e \geq 3v$.

c. For which of the Platonic solids is $2e$ exactly equal to $3v$?

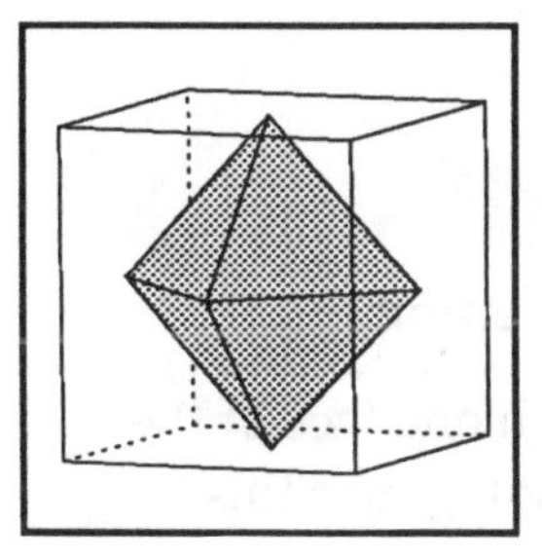

Activity 6: Duality of Polyhedra

1. a. How many faces does the cube have?

   How many vertices does the octahedron have?

   b. How many faces does the octahedron have?

   How many vertices does the cube have?

A polyhedron is ***inscribed*** *in a solid if it is inside the solid and each vertex of the polyhedron touches the surface of the solid.*

One polyhedron is the **dual** of another if its faces correspond in a special way to the vertices of the other and its vertices correspond similarly to the faces of the other. For example, the cube and the octahedron are duals of each other.

In the next exercise, you will demonstrate the dual relationship of the cube and the octahedron.

2. Open the transparent cube and place the octahedron inside so that each vertex of the octahedron touches the center of a face of the cube.

   a. Describe the position of the edges of the octahedron in relation to the cube.

   b. Tilt the cube so that you are looking directly down at one corner. Describe the position of the vertex at that corner in relation to the closest face of the octahedron inside.

One way to find the dual of a Platonic solid or an Archimedean solid is to connect the centers of adjacent faces of the solid. This produces a model of its dual inscribed in the polyhedron, just like your octahedron inside the transparent cube.

The dual of the dual of a polyhedron is the original polyhedron; that is, if one polyhedron is the dual of a second polyhedron, then the second is the dual of the first. Each Platonic solid is the dual of another Platonic solid.

3. Let's try this on the octahedron.

   a. How many faces meet at each vertex of the octahedron?

   b. One face of the dual of the octahedron is made by connecting the centers of adjacent faces that surround a vertex of the octahedron. Use segments to connect the centers which have been marked in the picture. *Connetc the centers of adjacent faces only.* What shape have you made?

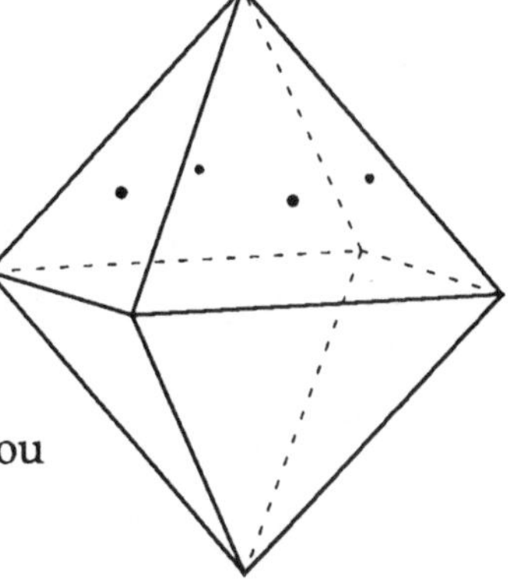

   c. On the picture of the octahedron, draw the rest of its inscribed dual.

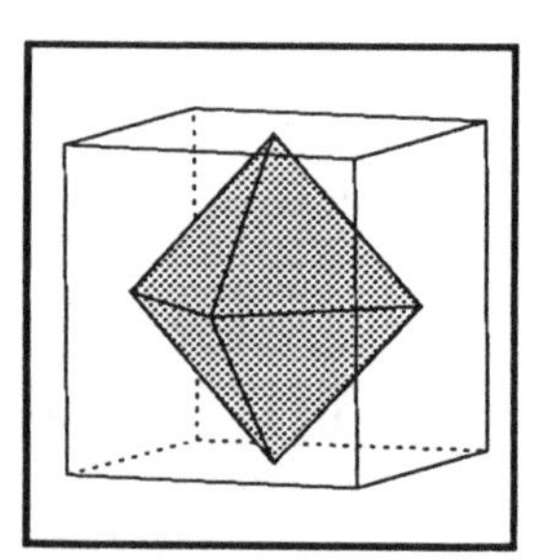

Activity 6 (continued)

4. The picture below shows one corner of a regular dodecahedron.

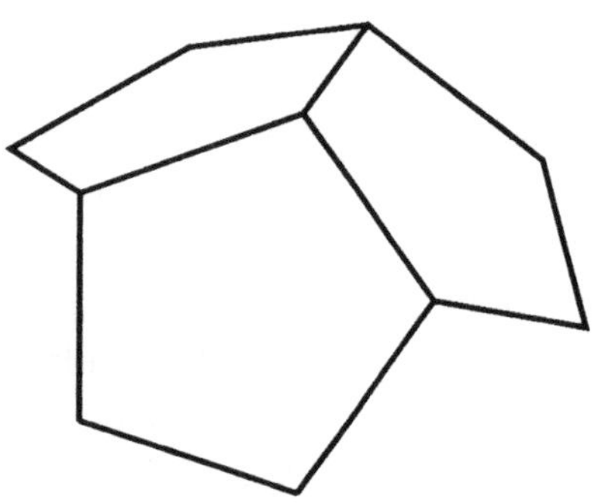

   a. On the picture, sketch one face of the dual figure. What polygon is this?

   b. How many vertices does the dodecahedron have?

   c. How many faces will the dual have?

   d. What polyhedron is the dual of a regular dodecahedron?

5. What polyhedron is the dual of a regular icosahedron? Explain.

6. What can you say about the number of edges of a polyhedron and its dual?

Archimedean solids have two or more kinds of regular polygon faces, but have identical vertices. Duals of Archimedean solids have congruent faces (why?) but are not regular polyhedra (why not?).

7. Draw the inscribed dual of the tetrahedron on the picture. What polyhedron is the dual of the regular tetrahedron?

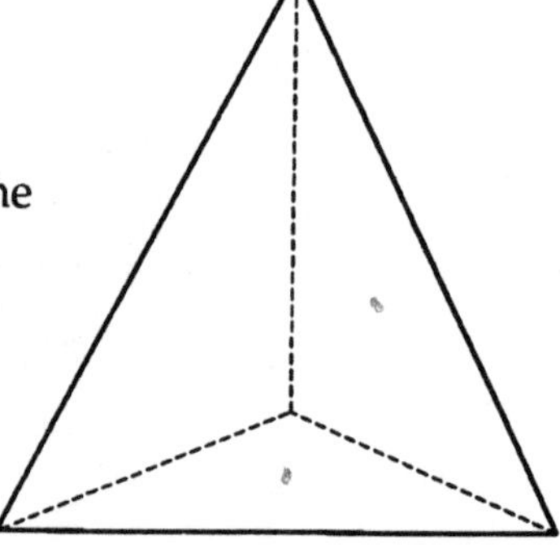

8. a. Examine the second chart in Activity 4 to find a pair of non-regular polyhedra that might be duals of each other. How can you test that the two polyhedra are duals?

   b. Find a polyhedron that is not regular and is dual to itself; that is, the shape of its dual is the same as the shape of the polyhedron.

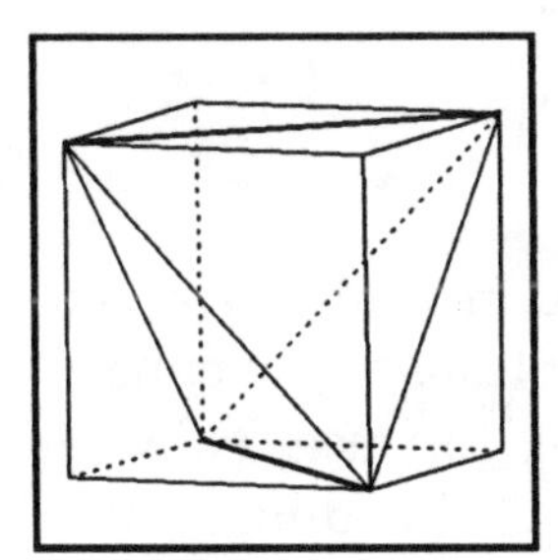

Activity 7: Inscribing Polyhedra in Each Other

A polyhedron is **inscribed** in a solid if it is inside the solid and each vertex of the polyhedron touches the surface of the solid.

Each of the Platonic and Archimedean solids can be inccribed in a sphere.

1. To inscribe a polyhedron in a solid, certain points on the surface of the solid are connected by line segments. For example, we can inscribe a cube in a dodecahedron by connecting a pair of non-adjacent vertices on each face of the dodecahedron.

   a. How many faces does the dodecahedron have?

   b. How many edges does the cube have?

   c. Each edge of the inscribed cube is a diagonal of one pentagon face of the dodecahedron. How many different diagonals does one face of the dodecahedron have?

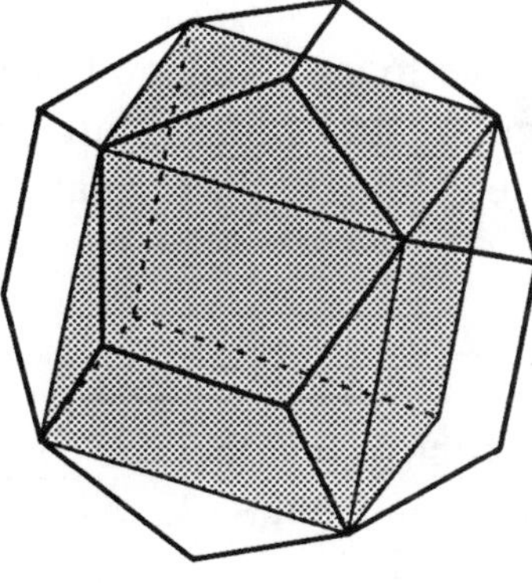

A compound of five interpenetrating cubes inscribed in a single dodecahedron is a beautifully symmetric star-like polyhedron.

   d. In how many different positions can a cube be inscribed in a dodecahedron?

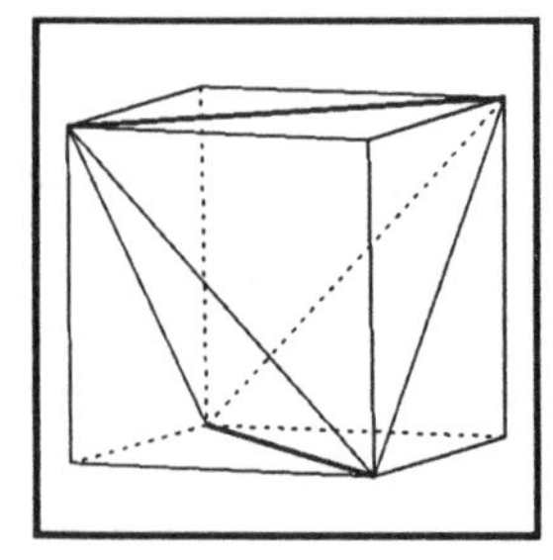

Activity 7 (continued)

2. In a similar fashion, a regular tetrahedron can be inscribed in a cube. You can see this by connecting certain non-adjacent vertices on the transparent cube. Use a non-permanent marker. First draw a diagonal on the top face. Now draw the *other* (non-parallel) diagonal on the bottom face. The endpoints of these two diagonals determine the four vertices of the tetrahedron. Connect the four vertices with line segments. Each edge of the tetrahedron will be a diagonal of a face of the cube.

   a. Where are the vertices of the tetrahedron in relation to the cube?

Since a regular tetrahedron can be inscribed in a cube, and a cube can be inscribed in a regular dodecahedron, imagine how a regular tetrahedron could be inscribed in a regular dodecahedron. The vertices of the tetrahedron will touch the vertices of the dodecahedron.

   b. In how many different positions can a tetrahedron be inscribed in a single cube in this way?

3. Estimate the volume of the inscribed tetrahedron compared to the volume of the cube.

4. (*optional*) Check your estimate by building a regular tetrahedron and four "cube corners" (net page 7).

   a. Put two cube corners together to form a pyramid with the same base and the same height as the regular tetrahedron. How does the volume of one cube corner compare to the volume of the regular tetrahedron?

   b. Fit all five models into the cube so that they are completely contained. Do you still believe your estimate in Exercise 3? Why or why not?

Activity 7 (continued)

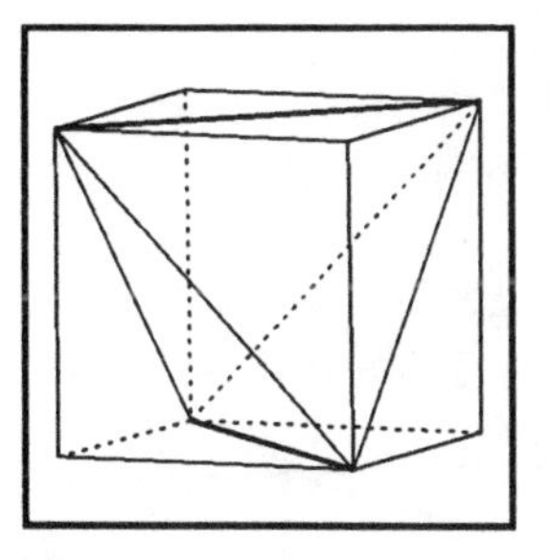

The cuboctahedron can be described as a cube whose corners have been removed by slicing through the midpoints of adjacent edges. The cuboctahedron can also be thought of as an octahedron whose corners have been sliced off in the same manner.

5. Wipe off the marks you have made on your transparent cube. Now draw line segments connecting the midpoints of *adjacent* edges of the cube. The inscribed solid you have created, which is called a cuboctahedron, has both square and triangular faces.

   a. How many square faces does it have?

   Where are the square faces of the cuboctahedron positioned in relation to the cube?

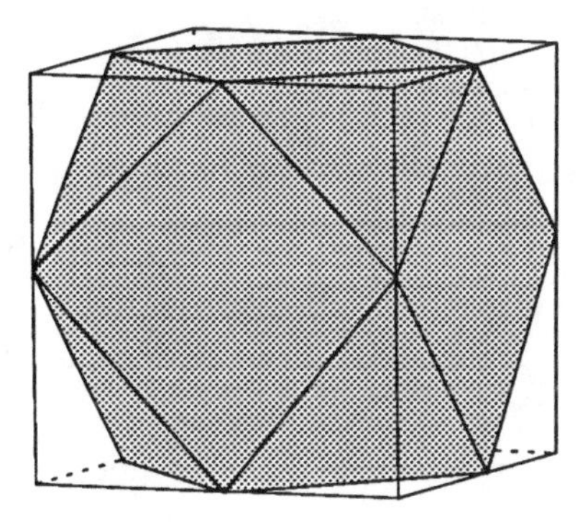

   b. How many triangular faces does the cuboctahedron have?

   Where are the triangular faces of the cuboctahedron positioned in relation to the cube?

6. A cuboctahedron can also be inscribed in a regular octahedron by connecting the midpoints of adjacent edges of the octahedron.

   a. Where will the square faces of the cuboctahedron be in relation to the octahedron?

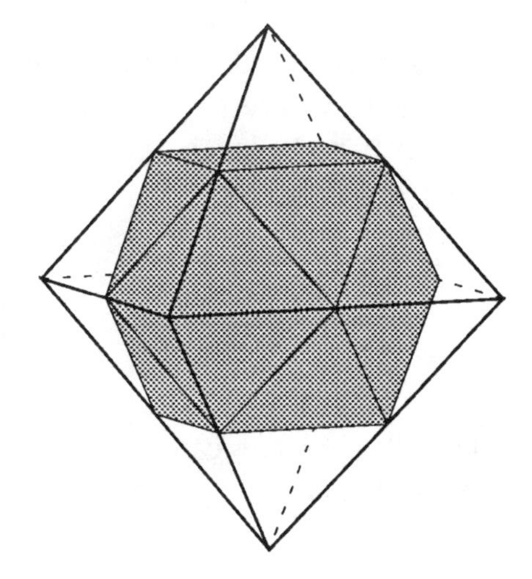

   b. Where will the triangular faces of the cuboctahedron be in relation to the octahedron?

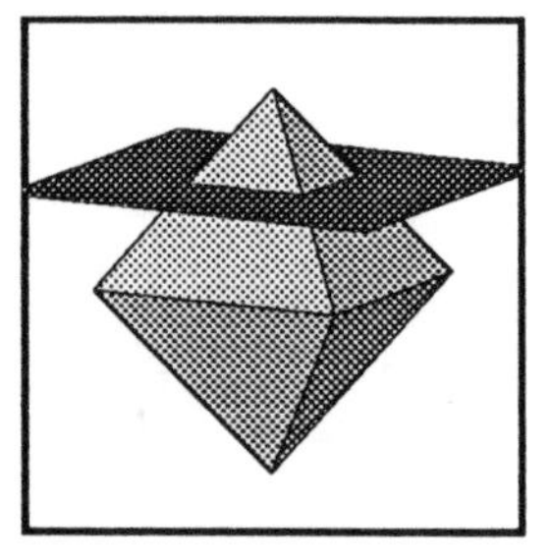

Activity 8: Cross Sections

1. Use the transparent cube. Wrap a rubber band *across* 4 edges of the cube, as shown. It does not have to bisect the edges of the cube, but it should be perpendicular to each edge of the cube it crosses.

   a. How many faces of the cube are crossed by the rubber band?

   Imagine a plane slicing the polyhedron into two pieces. The cross section is the new face produced from the slice.

   b. What shape is outlined by the rubber band?

Definition: A **cross section** of a polyhedron is the polygon produced when a plane intersects the polyhedron.

2. The rubber band in Exercise 1 represents a plane slicing through the cube, parallel to two faces of the cube. In this example, a square cross section is produced.

   a. Consider some other planes that form square cross sections when they intersect the cube. Are all these square cross sections of the cube congruent?

3. Change the location of the rubber band so that it forms a cross section that is rectangular, but not square.

   a. Are all rectangular cross sections of the cube congruent?

   b. If you wanted to produce a rectangular cross section with the largest area possible, where should you put the rubber band? Sketch the rubber band on the picture.

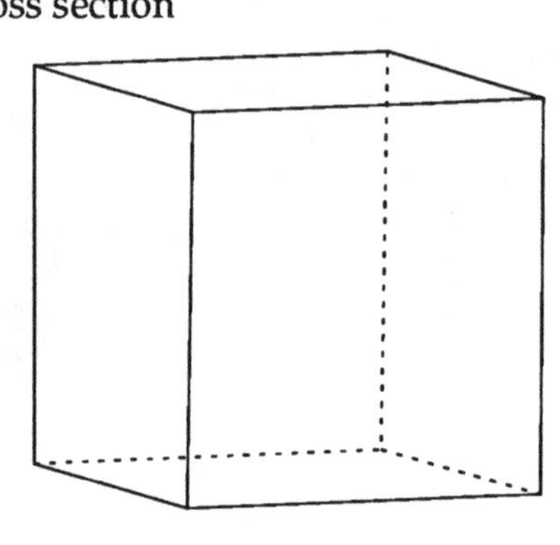

   c. How many of these largest rectangles can appear as cross sections of the cube?

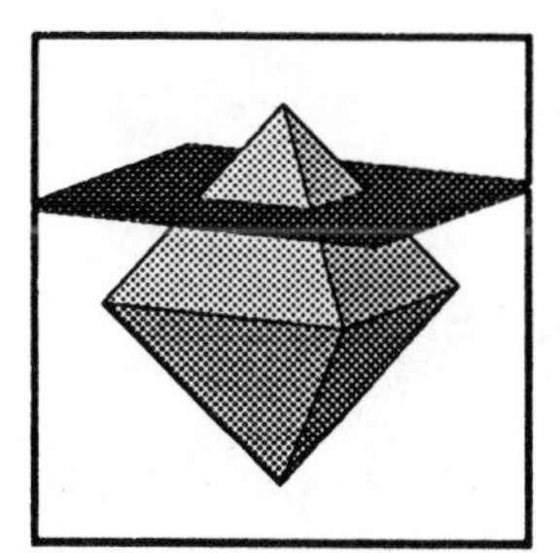

Activity 8 (continued)

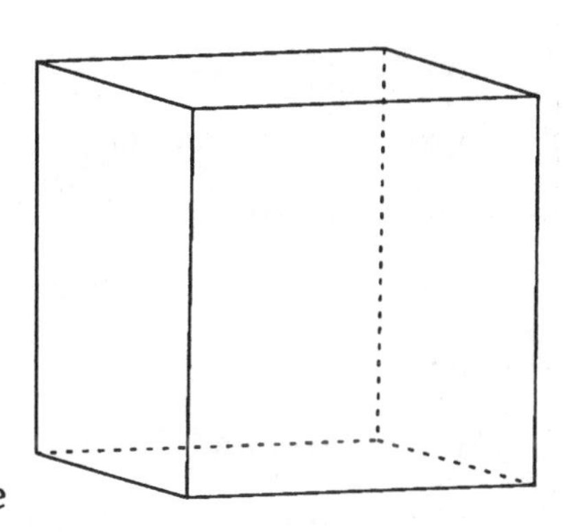

4. Experiment with your cube and rubber band to find examples of other quadrilateral cross sections, such as a rhombus (but not a square) or a parallelogram (but not a rhombus or rectangle). Sketch your results, labeled appropriately, on the drawing of the cube.

5. a. Describe how to slice the cube to form an equilateral triangle as a cross section of the cube. (Where are its vertices?)

   b. Describe a largest equilateral triangle which can be obtained as a cross section of the cube. (Where are its vertices?)

   c. How many of these largest equilateral triangles can appear as cross sections of the cube?

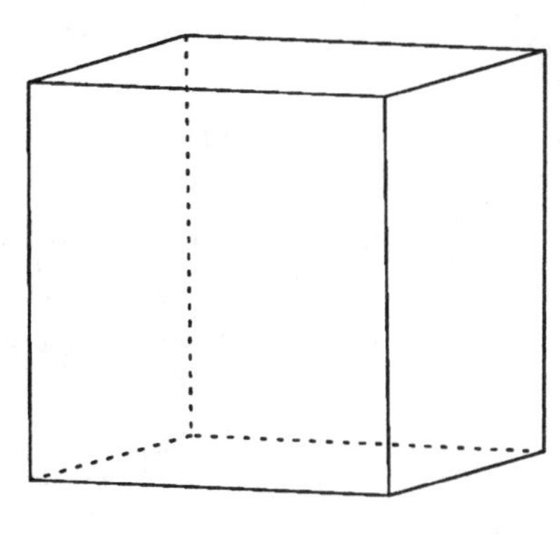

6. Experiment with your cube and rubber band to find examples of other triangular cross sections, such as an isosceles triangle or scalene triangle. Is it possible to get a right triangle or an obtuse triangle?

   Sketch your results, labeled appropriately, on the drawing to the left.

Can you find a pentagonal cross section of the cube? Is it possible to find a cross section of the cube which is a regular *pentagon?*

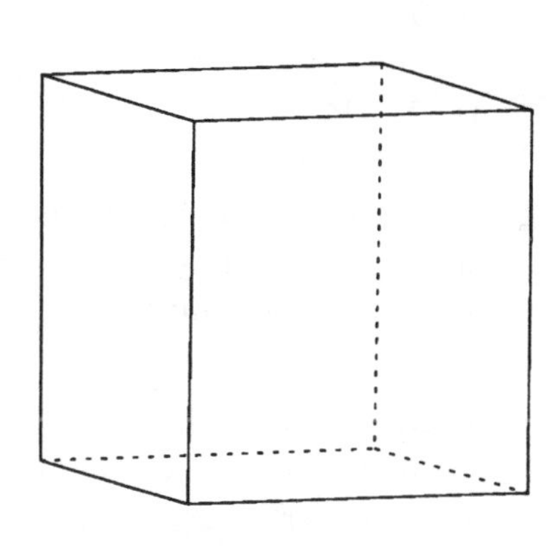

7. a. How many faces of the cube must be cut by a plane to produce a hexagonal cross section?

   b. Place a rubber band on your cube to demonstrate a cross section which is a regular hexagon. Sketch the rubber band on the picture.

Student Project 1: Semiregular Tessellations

Materials

Several triangles, squares, hexagons, octagons, and dodecagons (net pages SP-1A and SP-1B).

Background Knowledge

Activities P-1 and P-3.

In this project, you will determine how many different semiregular tessellations are possible.

Review

1. How do you find the measure of each interior angle of a regular polygon with n sides?

2. What's the difference between a *regular* tessellation and a *semiregular* tessellation?

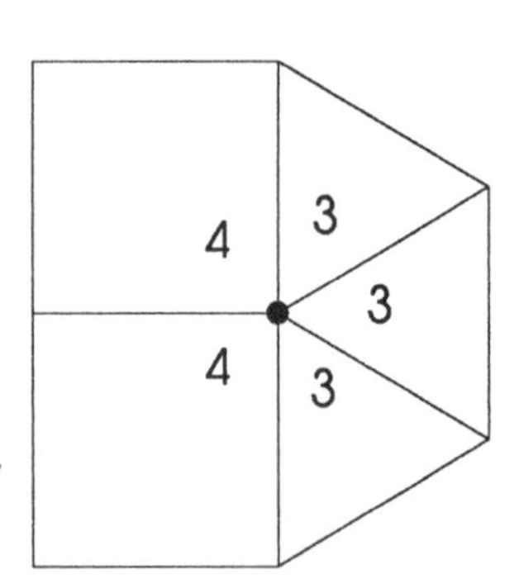

3. You can arrange three triangles and two squares around a vertex with no gaps, since the total angle measure around the vertex will be $3(60°) + 2(90°) = 360°$. The code 3-3-3-4-4 can be used to describe the arrangement here (the numbers tell how many sides each shape has; they are ordered according to the arrangement of the shapes around the vertex). How else might three triangles and two squares be arranged around a vertex? (Use a code like the one above to describe this arrangement.)

Experiment

4. Make several copies of the polygons on net pages SP-1A and SP-1B and cut them out. What arrangements of polygons (using more than one kind) can you fit around a vertex without any gaps? List as many as you can (you should find 12), using a code to describe them. (Be sure to consider different orders for each set of shapes.)

5. Now, for each arrangement in Exercise 4, try to continue the tessellation by adding more of the same kinds of polygons. Be sure that the arrangement of polygons around each vertex is the same. A few of them cannot be continued this way. See if you can find these and explain why they don't continue to form a semiregular tessellation.

6. Here are codes for some other arrangements of two or more regular polygons about a vertex. For each one, show that the total angle measure is 360°.

   a. 5-5-10 b. 4-5-20 c. 3-7-42

   d. 3-8-24 e. 3-9-18 f. 3-10-15

7. Without making all the regular polygons needed to try out the arrangements in Exercise 6 above, we can argue that none of these arrangements will form a semiregular tessellation. For example, both 6a and 6b involve a pentagon surrounded by two non-congruent polygons. That is, each vertex is surrounded by three angles—one measuring 108° (from the pentagon) and two others (a and b) which are not equal to each other. But that arrangement cannot be continued, as the following diagram illustrates.

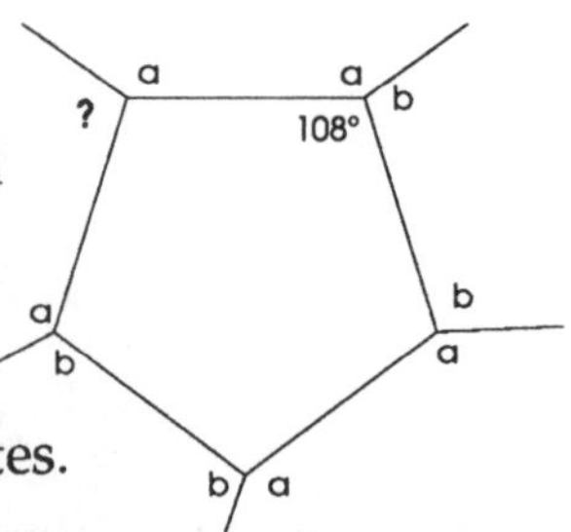

Student Project 1 (continued)

Therefore, neither 6a nor 6b can be continued to form a semiregular tessellation.

Each of the remaining four cases in Exercise 6 has a triangle surrounded by two non-congruent polygons. Explain why no tessellation is possible in these cases.

Conclusion

8. How many different semiregular tessellations are there altogether?

References

J. Britton and D. Seymour, *Introduction to Tessellations,* Chapter 3 and Appendix

K. Critchlow, *Order in Space*

B. Grünbaum and G. Shephard, *Tilings and Patterns an Introduction,* Chapter 2

P. O'Daffer and S. Clemens, *Geometry: An Investigative Approach,* Chapter 3

Student Project 2: Rotation Symmetries of the Regular Dodecahedron and the Regular Icosahedron

Materials

Regular dodecahedron and icosahedron from Activity 1, straws or pipe cleaners.

Background Knowledge

Activity 2.

Review

1. What does it mean for a polyhedron to have n-fold rotation symmetry?

Experiment

2. Pick up your dodecahedron and put a straw or pipe cleaner through a pair of opposite vertices. What kind (?-fold) of rotation symmetry does the dodecahedron have around this axis?

   How many such axes does the dodecahedron have?

3. Now put a straw or pipe cleaner through the centers of two opposite faces. What kind of rotation symmetry does this demonstrate?

   How many such axes does the dodecahedron have?

4. Finally, put a straw through the midpoints of two opposite edges, passing through the center of the polyhedron. What kind of rotation symmetry does this demonstrate?

   How many such axes does the dodecahedron have?

5. Repeat your investigations on the regular icosahedron. Summarize your results for the dodecahedron and the icosahedron in a table similar to the one in Activity 2, Exercise 8. Compare the rotation symmetries of the two shapes.

Student Project 3: Reflection Symmetry of the Platonic Solids

Materials

Seeds for the five Platonic solids (net pages SP-3A through SP-3E); three rectangular mirrors, at least 6 inches by 6 inches.

Background Knowledge

Activity 2.

Review

1. Describe the reflection planes of a cube.

Experiment

Lay one mirror flat, with the reflective side up. Place the other two mirrors upright, at a 45° angle to each other. You may want to make a tape hinge for the two upright mirrors.

2. Build the cube seed. Place it so that the "MIRROR" faces are flat against the two standing mirrors. You should see the image of a cube in the mirror kaleidoscope. Explain this image in terms of the reflection planes of the cube. How many images of the cube seed make the whole cube?
3. Build the octahedron seed. Describe the seed itself. What shape is it?

   Place the seed in the mirror kaleidoscope to see the image of an octahedron.
4. For the remaining three Platonic solids, remove the bottom mirror. Build the seeds, and place them against the standing mirrors. Close the mirrors until the seed is held in place. What is the measure of the angle between the two standing mirrors:
   a. for the tetrahedron seed?
   b. for the dodecahedron seed?
   c. for the icosahedron seed?

Challenge

The smallest seed for each Platonic solid is formed when all of the reflection planes cut the Platonic solid into parts. Describe the minimum seed for each Platonic solid, and determine how many of its images make the whole solid.

References

Films: *Dihedral Kaleidoscopes*
Symmetries of the Cube
Materials: *Root Blocks* by Rhombics (these blocks are the smallest cube seed)

Student Project 4: The Truncated Octahedron

Materials

8 hexagon panels, 6 square panels, 12 triangle panels (net page SP-4), rubber bands.

Background Knowledge

Activities 1 and 3.

Review

An Archimedean solid is built from two or more different kinds of regular polygons, so that each vertex of the solid is exactly the same. See the **Glossary** for pictures of the 13 Archimedean solids.

Definition

Truncation of a polyhedron occurs when all its corners or edges are sliced off in the same manner.

In this project, you will build a truncated octahedron and three corners of the original octahedron to demonstrate the truncation.

Experiment

Build the truncated octahedron by attaching hexagons and squares so that there are two hexagons and one square surrounding each vertex. Stand the figure on one of its square faces.

1. Build three pyramid-like corners, each with four triangles meeting at a vertex. Place one corner on the square face at the top of the truncated octahedron, and hold the other two on the square faces that show in the picture above.

   If you attached one of these corners to *each* square face of the truncated octahedron, what shape would you have?

2. A truncated octahedron is formed by slicing the corners off a regular octahedron so that the faces of the new solid formed are regular polygons. Demonstrate this now by removing the corners formed from triangles.

   What shape is the new face that is created when one corner is sliced off?

3. The truncated octahedron is unique among the Archimedean solids in that several copies of this figure will pack together without any gaps.

   The angle between two faces of a polyhedron is called a **dihedral angle**. Look up the dihedral angles of the truncated octahedron in one of the references below, and show that there is a combination of these angles which has a sum equal to 360°. (This explains why the packing works.)

4. Each Platonic solid has a truncated form that is an Archimedean solid. For example, imagine truncating all the corners of a cube so that portions of the original edges remain and all the new faces are regular polygons.

   What shape is the new face that is created when one corner is sliced off?

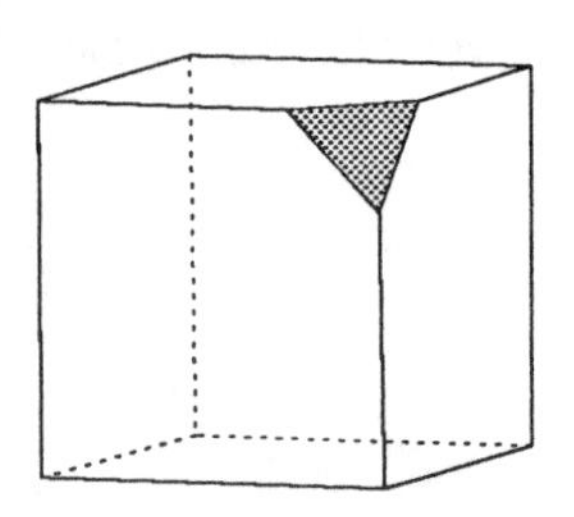

Student Project 4 (continued)

After all the corners have been truncated, what shape replaces each original square face?

5. Repeat Exercise 4 for the tetrahedron, dodecahedron, and icosahedron. For each truncated solid, how is the number of new faces related to the original Platonic solid?

   How is the shape of the new faces related to the original Platonic solid?

For more on the truncated icosahedron, see Student Project 9.

References

K. Critchlow, *Order in Space* (has measures of dihedral angles)

H. M. Cundy and A. Rollet, *Mathematical Models* (has measures of dihedral angles)

A. Holden, *Shapes, Space, and Symmetry*

P. O'Daffer and S. Clemens, *Geometry: An Investigative Approach,* Chapter 4

H. Steinhaus, *Mathematical Snapshots,* Chapters 7 and 8

Student Project 5: The Rhombic Dodecahedron

Materials

Cube pattern (net page SP-5A), 6 pyramids (net page SP-5B), transparent cube, tape.

Background Knowledge

Activity 3.

Experiment

This project relates the rhombic dodecahedron to other shapes, particularly the cube.

Cut out the cube pattern. Glue the two halves together as indicated. Score the lines.

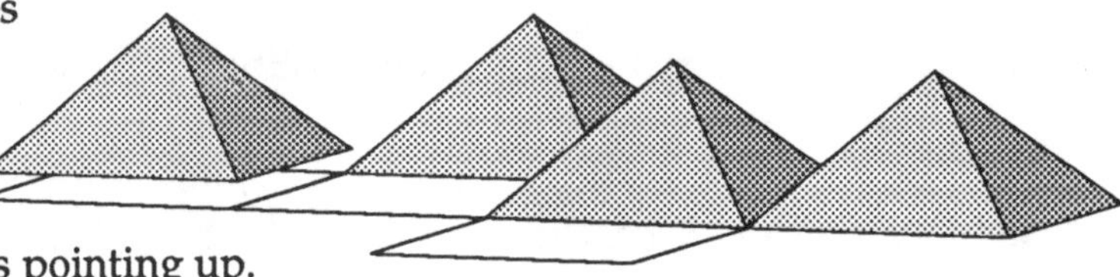

Build six pyramids, using net page SP-5B as a pattern. Glue one pyramid onto each square of the cube pattern, with the pyramids pointing up.

Fold the shape up to make a cube, with the pyramids on the inside. Your cube should be the same size as the transparent cube.

1. Now unfold the pyramids and wrap them around the transparent cube, with the base of a pyramid touching each face of the cube. What is the name of the solid formed on the outside?
2. Your transparent cube demonstrates a cube inscribed in a rhombic dodecahedron. How is each edge of the cube related to a face of the rhombic dodecahedron?
3. What is the volume of the rhombic dodecahedron compared to the volume of the cube inside?
4. Suppose you draw the long diagonals of each face of the rhombic dodecahedron. These would form the edges of what inscribed shape?

References

Arthur Loeb, "The Rhombic Dodecahedron and Its Relation to the Cube and the Octahedron," *Shaping Space*, ed. M. Senechal and G. Fleck

Ian Stewart, "How to Succeed in Stacking," *New Scientist*, vol. 131, no. 1777, July 13, 1991

Student Project 6: Dihedral Angles

Materials

Scientific calculator, nets for the regular tetrahedron and regular octahedron (net page SP-6-13).

Background Knowledge

Activity 1, definition of trigonometric functions, Pythagorean Theorem, Law of Cosines.

In this project, you will calculate the dihedral angles of some Platonic solids. You need to know some trigonometry; a scientific calculator is recommended, as well.

Definitions

The angle at which two faces of a polyhedron meet is called a **dihedral angle**.

$\cos \beta = \frac{b}{c}$

$\sin \beta = \frac{a}{c}$

Pythagorean Theorem: In a right triangle with legs a and b and hypotenuse c, $a^2 + b^2 = c^2$.

In a right triangle, the **cosine** of an angle β is defined to be the ratio of the length of the side adjacent to β to the length of the hypotenuse. The **sine** of β is the ratio of the length of the side opposite β to the length of the hypotenuse.

arcfunctions: If $x = \cos \beta$, then $\beta = \arccos x$. If $y = \sin \beta$, then $\beta = \arcsin y$.

Law of Cosines: In any triangle with legs a, b, c and corresponding opposite angles A, B, and C, the following three rules apply:

$$a^2 = b^2 + c^2 - 2bc \cos A$$
$$b^2 = a^2 + c^2 - 2ac \cos B$$
$$c^2 = a^2 + b^2 - 2ab \cos C$$

Experiment

In a Platonic solid, all dihedral angles have the same measure. The measure of the dihedral angle is a property of the solid.

Build the octahedron and tetrahedron by folding along the solid lines and gluing the tabs.

1. What is the measure of the dihedral angle of a cube?

2. Now consider the octahedron and the tetrahedron. Hold them together by matching one pair of faces. Notice that each of the other faces of the tetrahedron is coplanar with a face of the octahedron. This shows us that the dihedral angles of the two solids are supplementary. Because of this property, octahedra and tetrahedra together will pack space with no gaps. The framework of edges in this packing (called an octet truss) forms a lightweight but rigid structure, and so it is often used in architecture and scaffolding.

   Now we'll calculate the measures of these two angles.

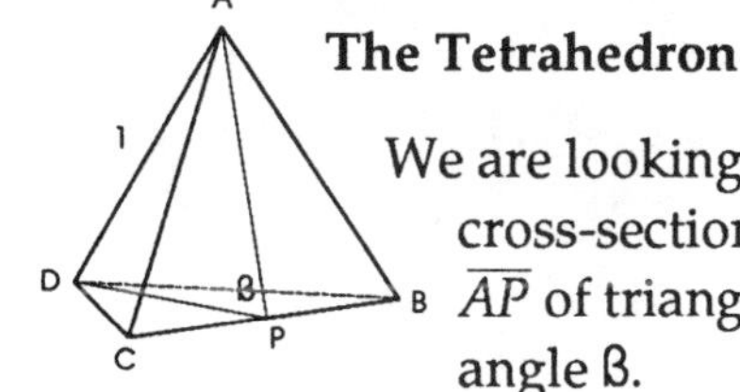

The Tetrahedron

We are looking for the dihedral angle between faces ABC and BCD. First we need a cross-section plane perpendicular to $\overline{BC}$. We form this plane by drawing the altitude $\overline{AP}$ of triangle ABC and the altitude $\overline{DP}$ of triangle DCB. Angle APD is the dihedral angle β.

Student Project 6 (continued)

To do the calculations, let's assume each edge of the tetrahedron is equal to 1.

3. Use the Pythagorean Theorem to show that $AP = \sqrt{3}/2 = DP$.
4. By the Law of Cosines, $AD^2 = AP^2 + DP^2 - 2(AP)(DP) \cos ß$. Show that $\cos ß = 1/3$.
5. Use a table or calculator to find an approximate value for the measure of ß.

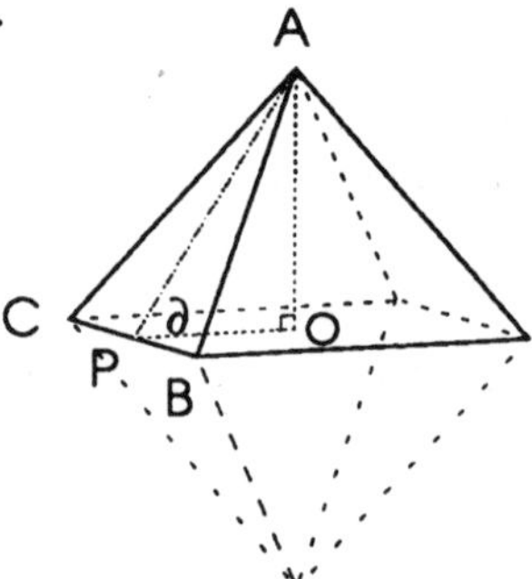

The Octahedron

Think of the octahedron as built from two regular square pyramids. The dihedral angle ∂ between the base and one face of the pyramid equals half the dihedral angle of the octahedron. We find this angle ∂ by drawing the altitude $\overline{AO}$ of the *pyramid* and the altitude $\overline{AP}$ of triangle *ABC*. The dihedral angle ∂ is angle *APO*.

6. Assume the length of each edge of the square pyramid is equal to 1. Then $\overline{OP} = 1/2$ and $\overline{AP} = \sqrt{3}/2$. Why?
7. Show that $\cos \partial = 1/\sqrt{3}$.
8. Use a table or calculator to find an approximate value for the measure of ∂.
9. The dihedral angle of a regular octahedron is 2∂. What is the measure of this angle?

Check that your results in Exercises 5 and 9 give supplementary angles.

Challenge

You can use trigonometric identities to prove that the sum of the dihedral angle of a tetrahedron and the dihedral angle of an octahedron is exactly 180°.

Begin with the results in Exercises 4 and 7: $\cos ß = 1/3$, $\cos \partial = 1/\sqrt{3}$.

9. Use the identity $\cos \partial = \sqrt{(1 + \cos 2\partial)/2}$ to show that $\cos 2\partial = -1/3$.

   This means that 2∂ and ß are supplementary. Why?

Dihedral Angles of Other Polyhedra

In an Archimedean solid, the dihedral angles need not have the same measure. In fact, the cuboctahedron and the icosidodecahedron are the only two Archimedean solids whose dihedral angles are all equal. (Why?)

11. Look up the dihedral angles of the truncated octahedron in a reference below, and show that there is a combination of these angles which has a sum equal to 360°. This shows that truncated octahedra will pack space with no gaps.
12. Archimedean duals have congruent faces and equal dihedral angles. Find an Archimedean dual that will pack space.

Student Project 6 (continued)

References

For dihedral angles of regular polyhedra, see H. S. M. Coxeter, *Introduction to Geometry*, Chapter 10, Section 4. This contains a general method of computing the dihedral angle ß of any Platonic solid and derives the following general formula:

For a Platonic solid whose faces have n sides, with r faces meeting at each vertex,
$\beta = 2 \arcsin [\cos (\pi/r)/\sin (\pi/n)]$.

Another method of calculating dihedral angles for Platonic solids is in Chapter 4 of P. O'Daffer and S. Clemens, *Geometry: An Investigative Approach*. Packing space with regular and semiregular polyhedra is also discussed.

For calculated values of dihedral angles of Platonic solids and other polyhedra, see H. M. Cundy and A. Rollett, *Mathematical Models* or K. Critchlow, *Order in Space*.

For more on space packing and the octet truss (invented by Buckminster Fuller as the basis of "spaceframes"), see J. Kappraff, *Connections*, Chapter 10, and A. Edmundson, *A Fuller Explanation*.

The space packing of octahedra and tetrahedra is pictured in H. Steinhaus, *Mathematical Snapshots*, and A. Holden, *Shapes, Space, and Symmetry*.

Student Project 7: Applications of Euler's Formula

Materials

None.

Background Knowledge

Activity 5, algebra of inequalities.

Euler's Formula is extremely useful in analyzing convex polyhedra, as you'll see in this project.

Review

Euler's Formula ($f + v = e + 2$) is a relationship which relates the number of faces, vertices, and edges of any convex polyhedron.

Experiment

In Activity 5, you demonstrated that the relationships below hold for all polyhedra:

$$2e \geq 3f \qquad\qquad 2e \geq 3v$$

1. Combine these inequalities with Euler's Formula to show that every convex polyhedron has at least six edges.
2. Show that every convex polyhedron has at least four vertices and at least four faces.
3. Show that there is no convex polyhedron with exactly seven edges. (Hint: Suppose $e = 7$ for some convex polyhedron. Then use the inequalities above to determine restrictions on the number of faces and vertices the polyhedron might have. Show that this violates Euler's Formula.)

   Student Projects 9 and 11 have other applications of Euler's Formula. Still more may be found in the references below.

References

A. Beck, M. Bleicher, and D. Crowe, *Excursions Into Mathematics,* Chapter 1

J. Pedersen and P. Hilton, *Build Your Own Polyhedra,* Section 12.3

H. Rademacher and O. Toeplitz, *The Enjoyment of Mathematics,* Chapters 12 and 13

Student Project 8: Deltahedra

Materials

58 triangle panels, rubber bands.

Background Knowledge

Activity 1, Activity 5.

Definition

A **deltahedron** is a convex polyhedron all of whose faces are congruent equilateral triangles.

Aside from the Platonic solids, the deltahedra are the only convex polyhedra having all faces congruent regular polygons.

In this project, you will use Euler's Formula ($f + v = e + 2$) to determine how many deltahedra can be built.

Experiment

You already know three deltahedra—the regular tetrahedron, the regular octahedron, and the regular icosahedron. But more are possible, since not all vertices have to be alike.

Let v_3 be the number of vertices at which 3 faces meet.

Let v_4 be the number of vertices at which 4 faces meet.

Let v_5 be the number of vertices at which 5 faces meet.

(Why is there no need for a variable v_6?)

Explain the reasoning behind the following equations:

1. $v = v_3 + v_4 + v_5$
2. $3v_3 + 4v_4 + 5v_5 = 2e$
3. $3v_3 + 4v_4 + 5v_5 = 3f$

4. Now combine these three equations with Euler's Formula to show that the following relationship must hold: $3v_3 + 2v_4 + v_5 = 12$

This equation has 19 solutions. Some of these are indicated in the chart on the next page.

5. Fill in the rest of the chart with values of v_3, v_4, and v_5 that satisfy the equation.

Although there are 19 solutions to the equation in Exercise 4, only eight of these deltahedra can actually be built. (This was proved in 1947 by H. Freudenthal and B. L. van der Waerden.)

6. Try to build the models shown in your chart, using triangles and rubber bands from Activity 1. You will need 90 triangles (58 if you don't try to make the Platonic ones) to make all eight deltahedra. Try to explain why some of the cases don't produce deltahedra (remember that they must be convex).

Student Project 8 (continued)

Finally, for the models which can be built, write their names in the chart. Use the following list, which tells how many faces each deltahedron has:

Regular Tetrahedron (4 faces)
Triangular Dipyramid (6 faces)
Regular Octahedron (8 faces)
Pentagonal Dipyramid (10 faces)
Snub Dispheroid (also called a Siamese dodecahedron) (12 faces)
Triaugmented Triangular Prism (14 faces)
Gyroelongated Squares Pyramid (16 faces)
Regular Icosahedron (20 faces)

v_3	v_4	v_5	**Name**
0	0	12	Regular Icosahedron
0	1	10	
0	2		
0			
0			
0			
0	6	0	
1	0	9	
1	1		
1	2		
1			
1			
2	0		
2	1		
2			
2			
3	0		
3	1		
4	0	0	Regular Tetrahedron

References

A. Beck, M. Bleicher, and D. Crowe, *Excursions into Mathematics,* Chapter 1
J. Pedersen and P. Hilton, *Build Your Own Polyhedra,* pp. 77-80
B. M. Stewart, *Adventures Among the Toroids,* Chapter IV
Models: *Geometric Playthings,* J. and K. Pedersen

Student Project 9: The Truncated Icosahedron

Materials

12 pentagon panels, 20 hexagon panels, rubber bands.

Background Knowledge

Activity 1, Activity 4, Activity 5.

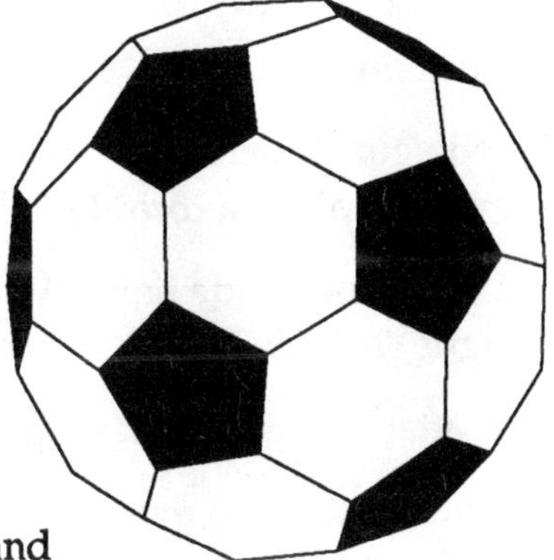

A standard soccer ball is formed from sewn leather polygons that are regular hexagons and pentagons. The soccer ball is made exactly like the Archimedean solid known as the **truncated icosahedron**.

The truncated icosahedron is the only possible semiregular polyhedron with regular hexagons and regular pentagons as faces. In this project, you will show why this is true.

Review

1. a. First, explain why any polyhedron built from regular hexagons and regular pentagons must have exactly three faces meeting at each vertex.

   b. Using part a, explain why the equation $2e = 3v$ must hold for any such polyhedron.

Experiment

2. Use Euler's Formula ($f + v = e + 2$) and the fact that $2e = 3v$ to show that any convex polyhedron built from regular hexagons and pentagons must have exactly 12 pentagonal faces. (Hint: Let x be the number of pentagonal faces, y the number of hexagonal faces. Find expressions for f, v, and e in terms of x and y. Substitute these into Euler's Formula, and solve for x.)

3. The only possible vertex configurations for a semiregular polyhedron built from regular hexagons and regular pentagons are (i) two hexagons and a pentagon, or (ii) two pentagons and a hexagon. Explain why the second option is impossible.

4. So far, you have shown that every vertex is surrounded by two hexagons and a pentagon, and that there are exactly 12 pentagon faces. Show that there must be exactly 20 hexagon faces. (Hint: every pentagon must be surrounded by five hexagons).

5. Build a truncated icosahedron, using 12 pentagon panels and 20 hexagon panels.

References

A. Beck, M. Bleicher, and D. Crowe, *Excursions Into Mathematics*, Chapter 1

R. Curl and R. Smalley, "Fullerenes," *Scientific American*, Oct. '91

Student Project 10: The Toroid

Materials

15 square panels, 6 triangle panels, rubber bands

Background Knowledge

Activity 1, Activity 5

Experiment

The polygon panels and rubber bands used in Activity 1 can be used to make a shape with a hole in it—an example of a **toroid**.

First, form the inside hole: Arrange 3 squares and 3 triangles as in the pattern at the right. Attach the end triangle to the end square to close the loop.

Now, attach three squares together in a row.

Then attach them to the squares on your loop. Close the top loop, also.

So far you have made the shape below.

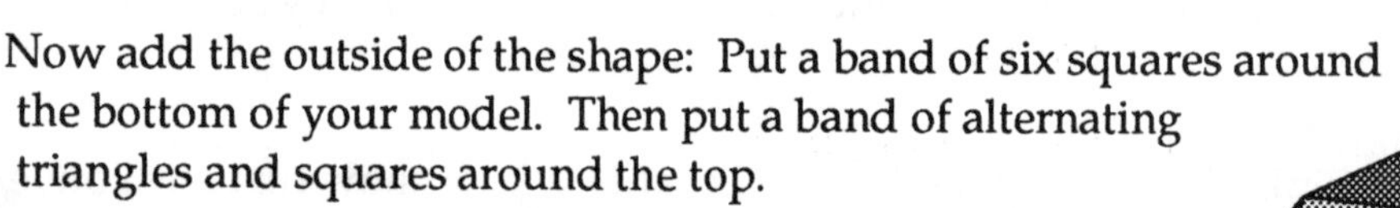

Now add the outside of the shape: Put a band of six squares around the bottom of your model. Then put a band of alternating triangles and squares around the top.

This completes the toroid. An interesting property of this shape is that it does not satisfy Euler's Formula ($f + v = e + 2$). Demonstrate this by counting the number of vertices, edges, and faces of the finished shape.

What *is* the relationship between f, v, and e for this shape?

Reference

B. M. Stewart, *Adventures Among the Toroids*, Chapter V

Student Project 11: Descartes' Theorem

Materials

None.

Background Knowledge

Activity P-1, Activity 5.

Review

The Exterior Angle Theorem says that for any convex polygon, the sum of the exterior angles is 360°.

Definition

Descartes' Theorem is the analogous statement about convex polyhedra. Instead of exterior angles, Descartes looked at the **defect** (or **deficit**) at each vertex of a convex polyhedron. To compute the defect at a vertex, sum all the face angles that meet at that vertex, and subtract the sum from 360°. The defect at a vertex gives a measure of how close the polyhedron is to being spherical.

Experiment

1. a. For each of the Platonic solids, find the defect at a vertex.

   b. Which solid has the smallest defect?

   c. Which solid seems to be the most round?

2. a. Find the sum of the defects for each Platonic solid. (Since the defect is the same at each vertex, you can just multiply the number of vertices of the solid by the defect at one vertex.)

   b. State a conjecture based on your findings:

To prove your conjecture, use the following variables:

f = number of faces
v = number of vertices
e = number of edges
n = number of sides on each face of the polyhedron
A = sum of all of the face angles of the polyhedron
d = defect at each vertex
D = sum of the defects

The key technique in the proof is to use two different ways to express A in terms of the other variables.

3. Assume you have a Platonic solid.

   a. First, we'll find an expression for A by adding all the interior angles of each polygon face.

   How many faces are there?

   What is the sum of the interior angles of each face?

   Complete: $A =$

   b. Now, we'll find A in terms of the defect. Explain why the following equation must hold: $A = v\,(360° - d)$.

   c. You have two equations involving A. Use the substitution principle to eliminate A. What equation does this give?

Student Project 11 (continued)

d. Now, use Euler's Formula ($v - e + f = 2$) and the facts that $fn = 2e$ (why?) and $vd = D$ (why?), to show that $D = 720°$.

4. a. Use the drawings of each of the Archimedean solids in the appendix to calculate the defect at each vertex and the sum of the defects for each Archimedean solid.

Name	Defect at Each Vertex	Sum of the Defects
Truncated Tetrahedron		
Truncated Cube		
Truncated Octahedron		
Truncated Dodecahedron		
Truncated Icosahedron		
Cuboctahedron		
Icosidodecahedron		
Snub Cuboctahedron		
Snub Icosidodecahedron		
Greater Rhombicuboctahedron		
Rhombicuboctahedron		
Greater Rhombicosidodecahedron		
Rhombicosidodecahedron		

b. Does your conjecture in Exercise 2b apply to the Archimedean solids?

Challenge

René Descartes (French mathematician, 1596-1650) proved that *for any convex polyhedron*, the sum of all the defects at its vertices is exactly 720°. A proof by mathematician George Pólya (1887-1985) is similar to the one you did in Exercise 3 for regular polyhedra. It is complicated by the fact that not all the faces are the same, so we let f_3 denote the number of three-sided faces, f_4 be the number of four-sided faces, and so on. Similarly, not all the vertices are the same, so we let a_n be the sum of the face angles about the vertex v_n. Try the proof yourself, or see the references.

References

J. Pedersen and P. Hilton, *Build Your Own Polyhedra*, pp. 159-161

D. Schattschneider, *Counting It Twice*

Student Project 12: Dual Polyhedra

Materials

Cube, six pyramids (net page SP-12).

Background Knowledge

Activities 3 and 6.

Review

The cube and the regular octahedron are dual polyhedra; that is, the faces of the cube correspond to the vertices of the octahedron, and the vertices of the cube correspond to the faces of the octahedron.

1. How does the number of *edges* of a cube compare to the number of *edges* of an octahedron?

Experiment

In this project, you will build a model showing a cube and an octahedron interpenetrated so that each edge of one of them is the perpendicular bisector of an edge of the other.

Build the cube first by folding along the solid lines and gluing the tabs. Copy the pyramid pattern on net page SP-12 so that you have six pyramid nets.

Color the triangular faces of the six pyramids all the same color, then build the pyramids. Glue the square face of each pyramid onto a face of the cube, as indicated by the dotted lines. The colored pyramids show the corners of an octahedron as though it interpenetrates the cube. (The video *The Platonic Solids* shows this as one view of the dual cube and octahedron.)

2. The two interpenetrating polyhedra share a shape as their inner core. What is the shape formed by the intersection of the cube and the octahedron?

3. Each edge of the cube is crossed by an edge of the octahedron; these are called **dual edges**. Each pair of dual edges forms the diagonals of what two-dimensional figure?

4. How many *pairs* of dual edges are there?

5. You should be convinced, from Exercises 3 and 4, that the smallest convex shape that will contain your model is a rhombic dodecahedron. How is this solid related to the inner solid in Exercise 2?

6. Other pairs of dual polyhedra can be arranged in this way (see A. Holden, *Shapes, Space, and Symmetry*, p. 9). In particular, a tetrahedron interpenetrated with its dual (another tetrahedron) forms a shape called the **stella octangula**, which is pictured here.

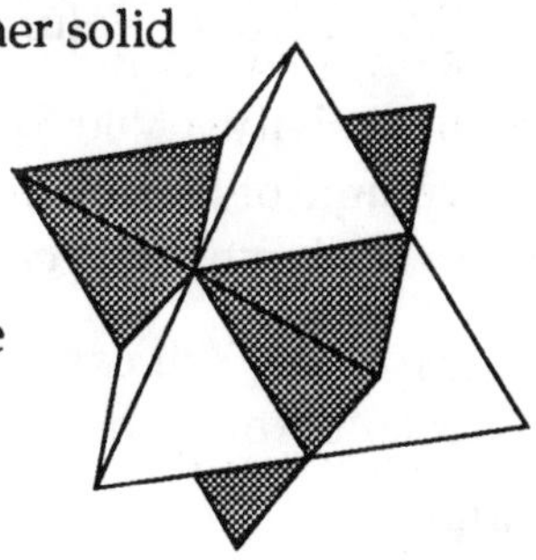

   What polyhedron is the inner core of the stella octangula?

   What is the smallest convex shape that will hold a stella octangula?

References

A. Holden, *Shapes, Space, and Symmetry*

Visual Geometry Project, *The Stella Octangula*

Visual Geometry Project, *The Platonic Solids*, Videotape

M. Wenninger, *Dual Models*

Student Project 13: The Volume of the Octahedron Inscribed in a Cube

Materials

Transparent cube, octahedron, 4 small tetrahedra, 4 "cube pieces" (net page SP-6 and 13).

Background Knowledge

Activities 6 and 7; volume of a pyramid.

Review

A regular octahedron can be inscribed in a cube so that each vertex of the octahedron touches the center of a face of the cube. This demonstrates the dual relationship of the octahedron and the cube.

Experiment

1. Build the octahedron (net page SP-6 and 13) and place it in the transparent cube. Estimate: How do you think the volume of the octahedron compares to the volume of the cube?

In this project, you will find the exact ratio for these volumes. To do this, you must build several small pieces to place in the cube with the octahedron.

2. First, build four cube pieces. Show how to assemble these four pieces (but don't glue them together) to make an octahedron congruent to the one in the cube. How does the volume of one cube piece compare to the volume of the octahedron?

3. Build one small tetrahedron (net page SP-6 and 13). Place the tetrahedron and one cube piece side by side, with the cube piece resting on its equilateral triangle face.

   Explain why the tetrahedron and the cube piece have the same volume.

   How does the volume of the tetrahedron compare to the volume of the octahedron?

4. Build three more small tetrahedra (net page SP-6 and 13). Glue the four small tetrahedra onto the faces of the octahedron to form a large tetrahedron. Show how the large tetrahedron fits in the transparent cube.

   How does the volume of the octahedron compare to the volume of the large tetrahedron?

5. In Activity 7, you found that the volume of a tetrahedron inscribed in a cube is equal to 1/3 the volume of the cube. Using this fact, explain how the volume of the inscribed *octahedron* compares to the volume of the cube.

6. (*optional*) You can make a cube puzzle by building 12 cube pieces, 8 small tetrahedra, and an octahedron. Challenge a friend to fit all these pieces into the cube.

References

M. Laycock and M. Smart, *Create a Cube*

Visual Geometry Project, *The Stella Octangula*

Student Project 14: The Tetrahedron Puzzle

Materials

Tetrahedron puzzle net (net page SP-14).

Background knowledge

Activity 8.

Experiment

1. Make two copies of the net, and build two identical models. Try to fit the two shapes together to form a regular tetrahedron. Describe the shape of the two faces that need to be matched to form the tetrahedron.
2. Describe precisely how to cut a tetrahedron into these two congruent pieces.
3. Challenge a friend to make the tetrahedron from the two shapes. What makes this puzzle so tricky is that the cross section plane that forms the two parts of the tetrahedron is not a reflection plane. Explain what you have to do to put the two shapes together to form a tetrahedron.

References

P. O'Daffer and S. Clemens, *Geometry: An Investigative Approach*, pp. 125-126 has three dissection puzzles.

M. Smart and M. Laycock, *Create a Cube*, has several dissection puzzles of the cube.

D. Stonerod, *Puzzles in Space*, has several dissection puzzles of the Platonic solids.

S.T. Coffin, *The Puzzling World of Polyhedral Dissections*, has several dissection puzzles of the Platonic solids.

Polygon Panels

Activity 1 and

Student Projects 4, 8, 9, and 10

Make seven copies of each of the two panels pages to make a complete set for activities. (Make additional copies for Student Projects, as necessary.)

Cut out the panels along the solid lines. Use a hole-punch to punch a hole or scissors to cut a narrow "V" at each vertex. Score and fold on dashed lines. If you use scissors to notch corners, make a narrow "V" as shown. If your "V" is too wide, your rubber bands won't stay on when you assemble your solids.

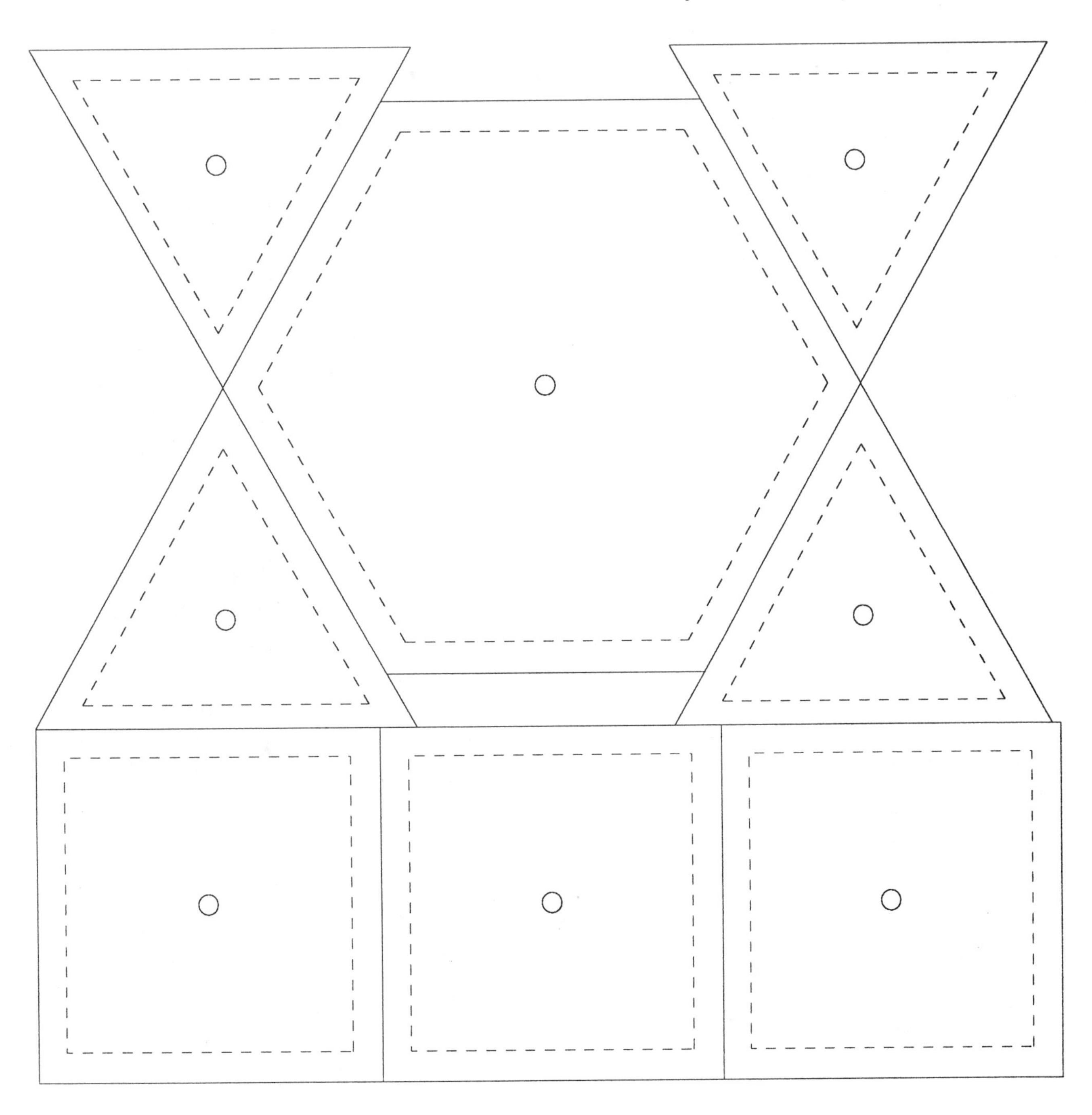

Polygon Panels

Activity 1 and

Student Projects 4, 8, 9, and 10

Make seven copies of each of the two panels pages to make a complete set for activities. (Make additional copies for Student Projects, as necessary.)

Cut out the panels along the solid lines. Use a hole-punch to punch a hole or scissors to cut a narrow "V" at each vertex. Score and fold on dashed lines.

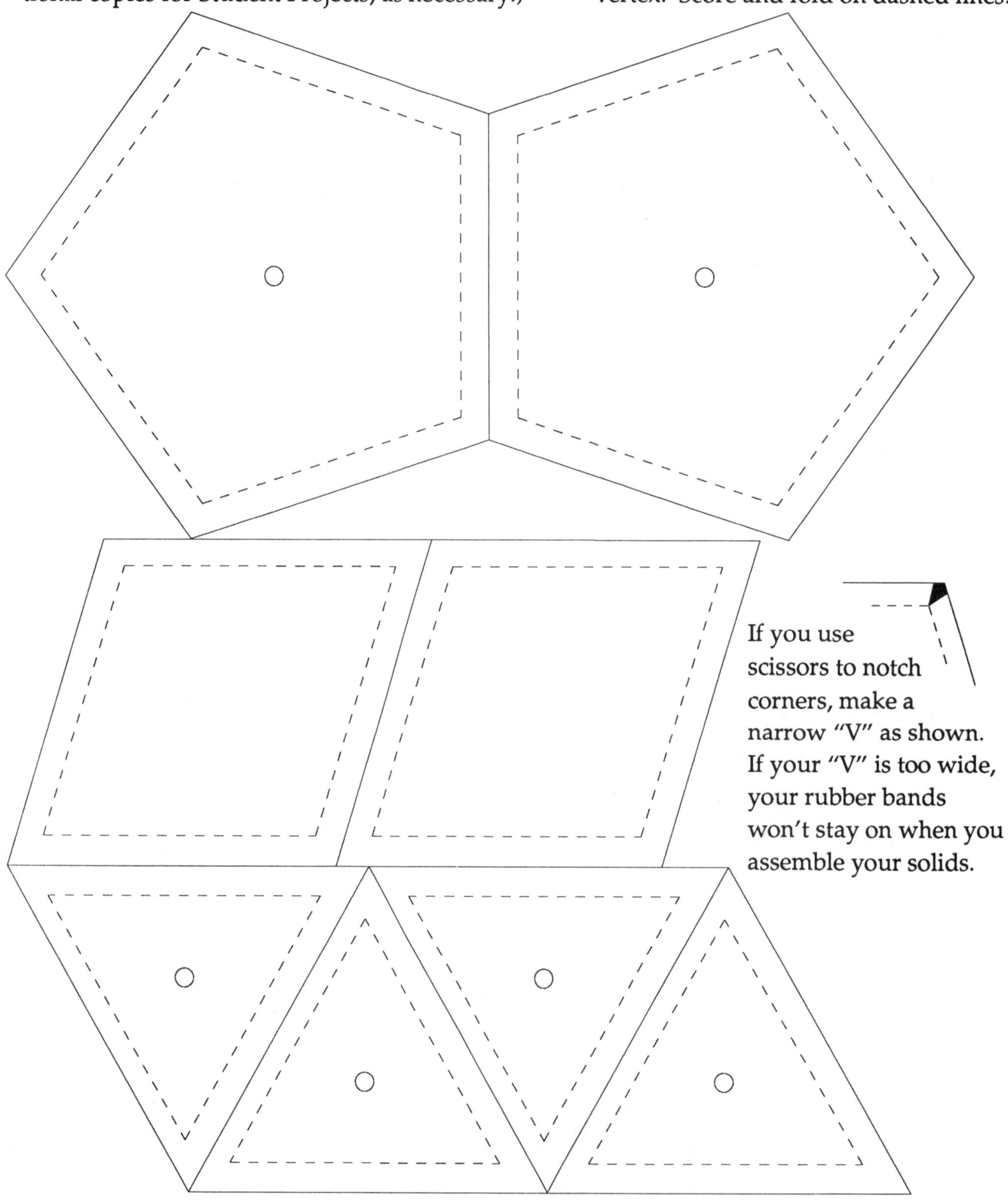

If you use scissors to notch corners, make a narrow "V" as shown. If your "V" is too wide, your rubber bands won't stay on when you assemble your solids.

Transparent Cube Net
(Transfer to 10 mil. acetate.)
Activity 2

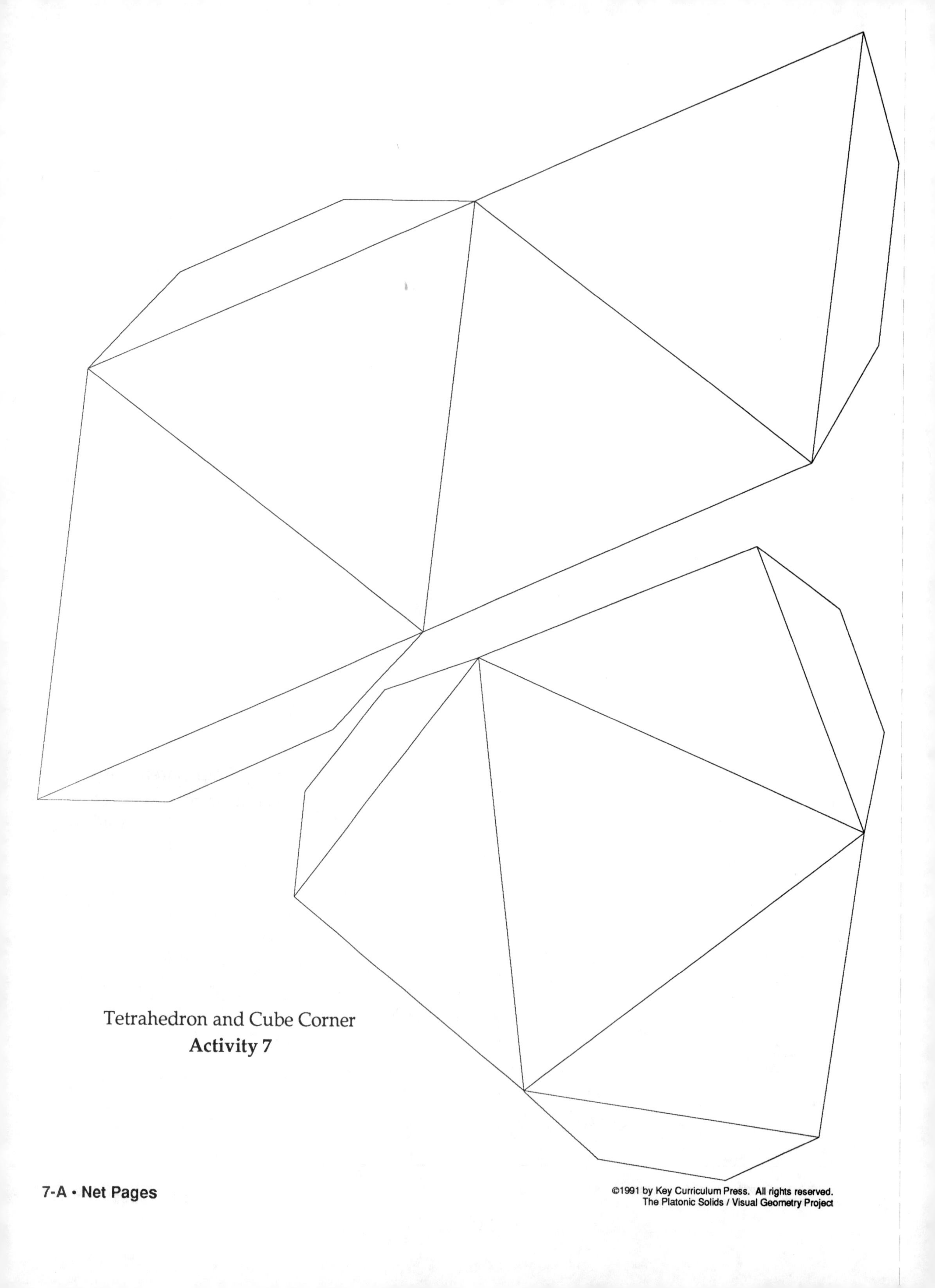
Tetrahedron and Cube Corner
Activity 7

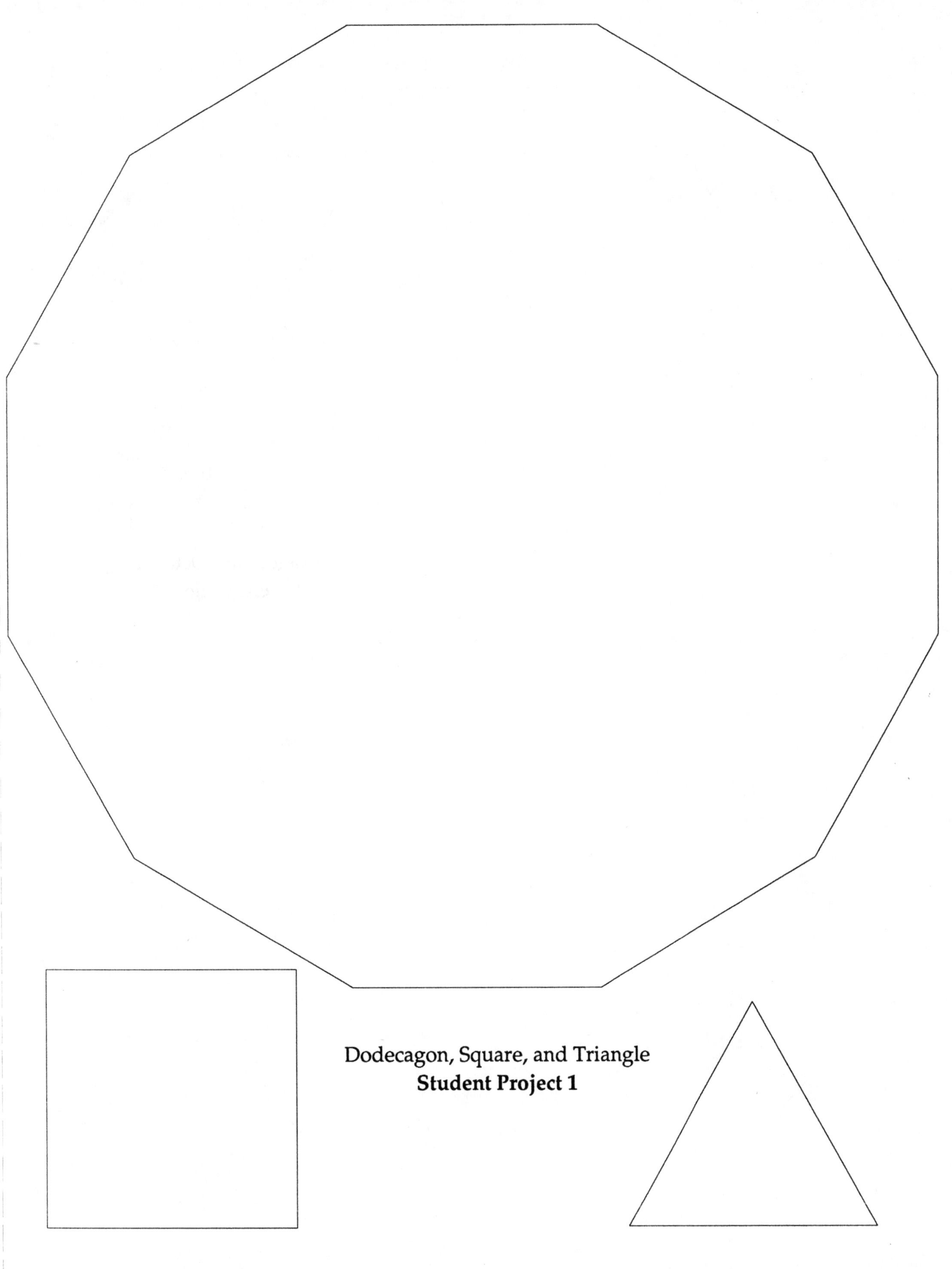

Dodecagon, Square, and Triangle
Student Project 1

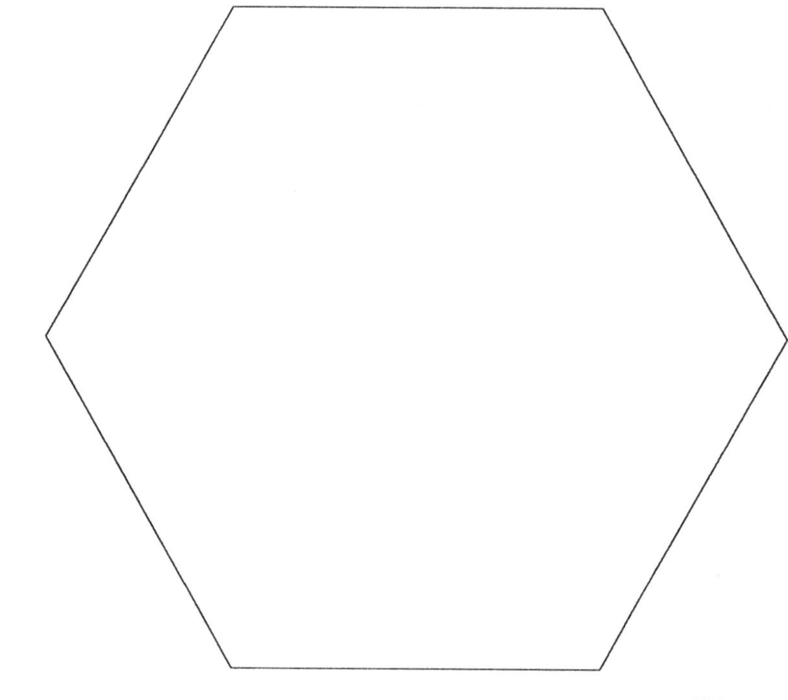

Hexagon and Octagon
Student Project 1

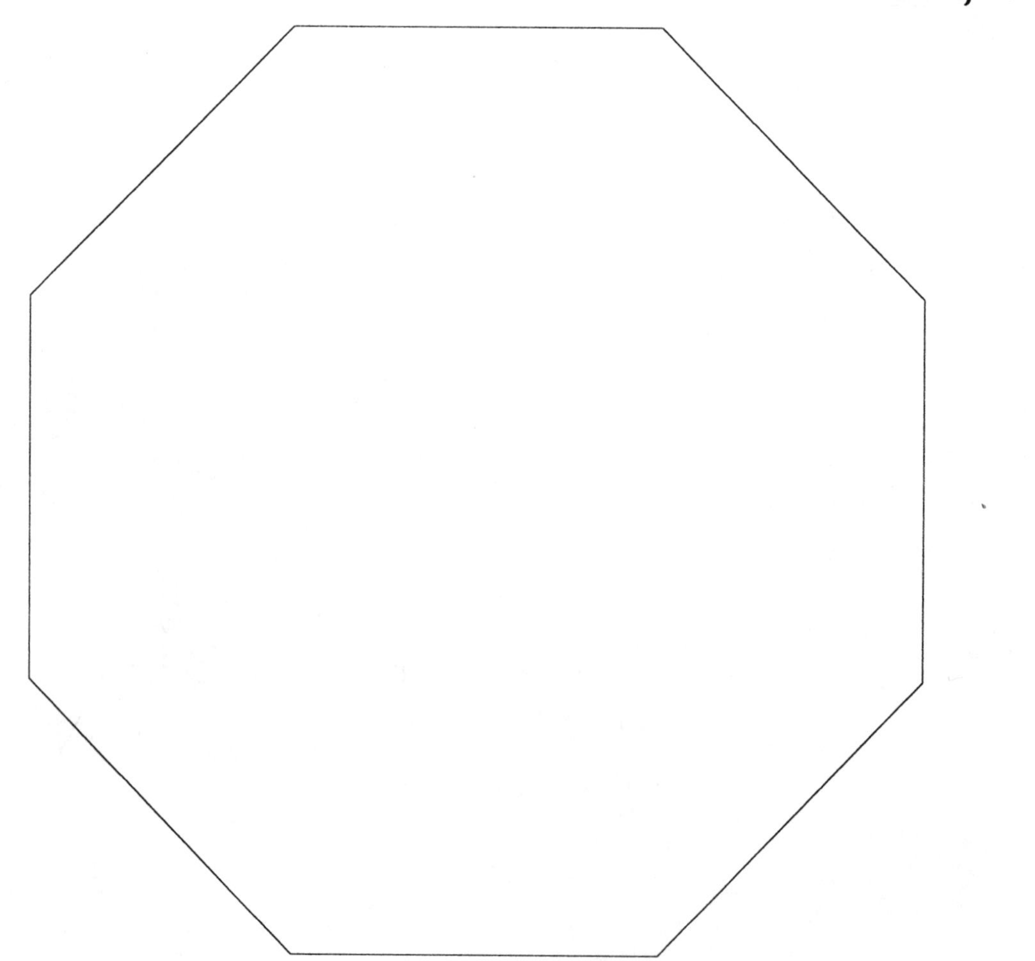

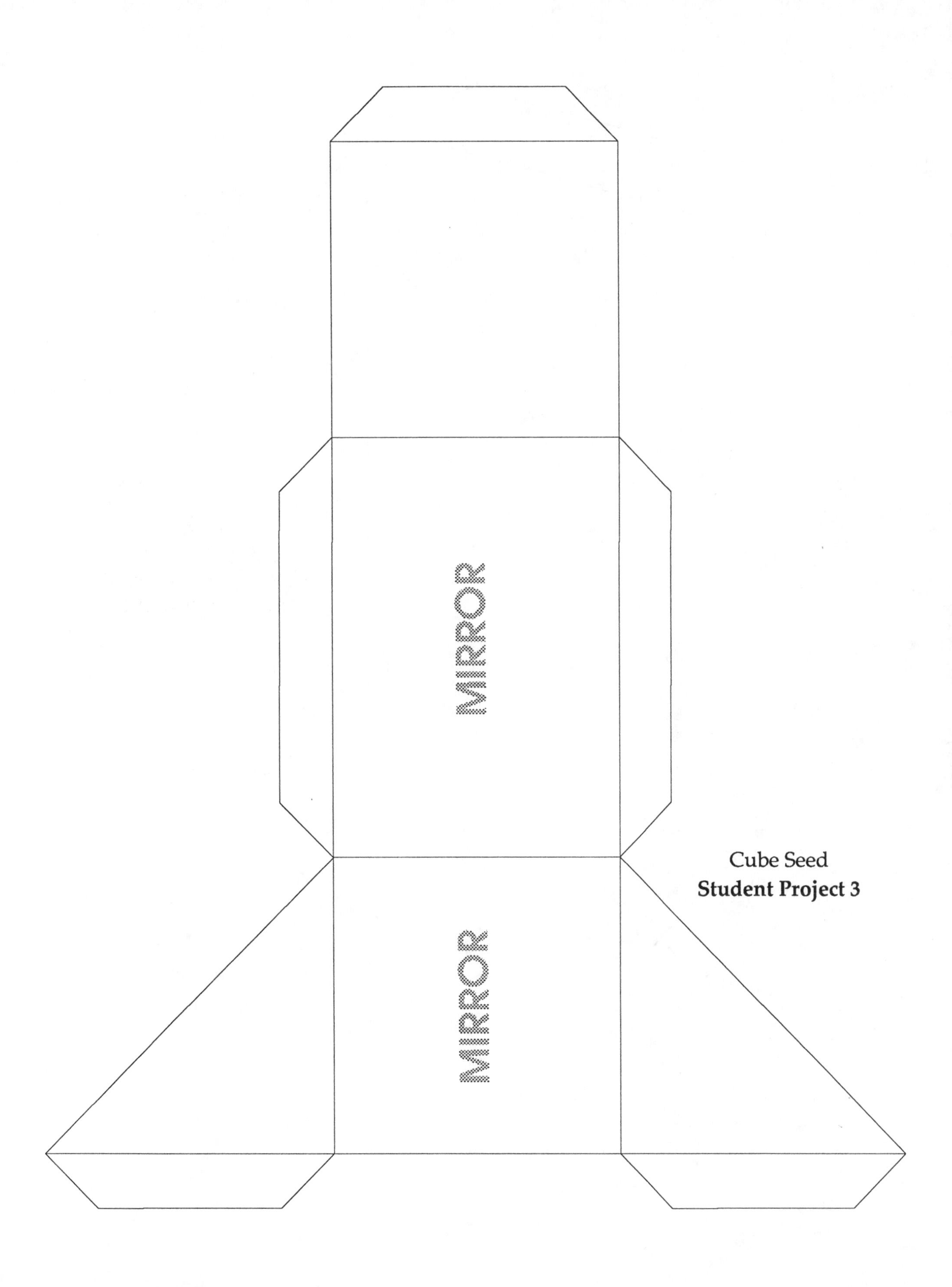

Cube Seed
Student Project 3

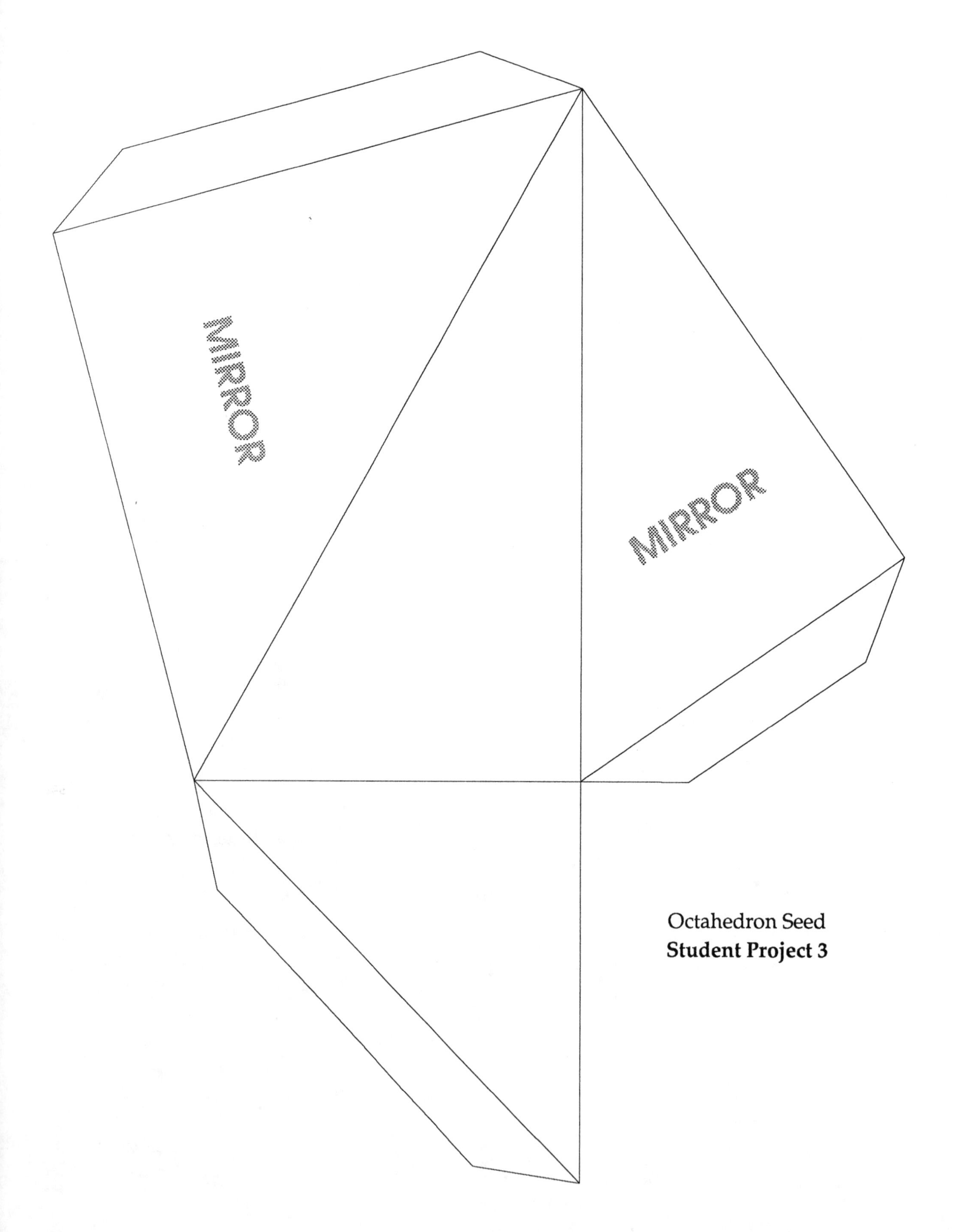

Octahedron Seed
Student Project 3

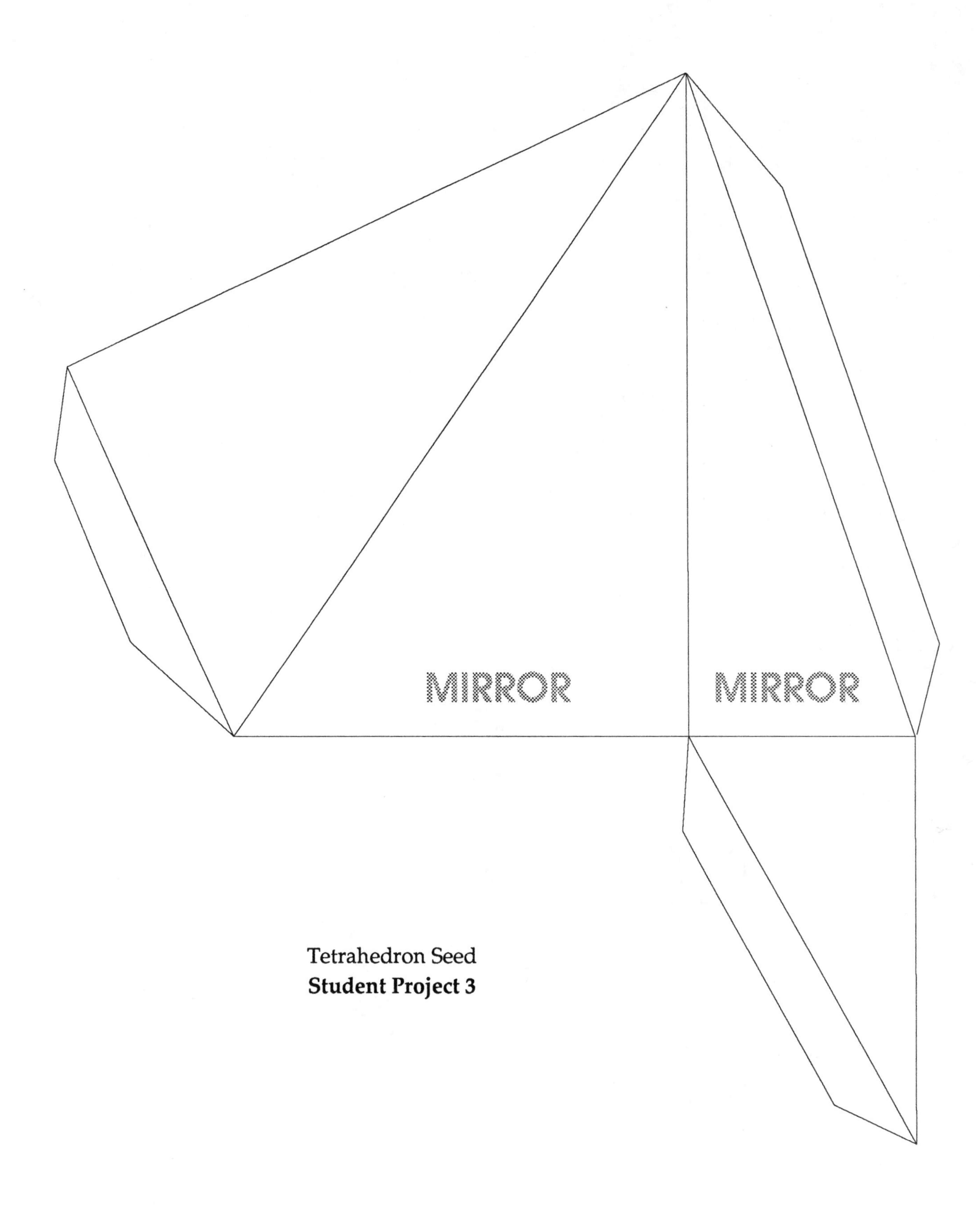

Tetrahedron Seed
Student Project 3

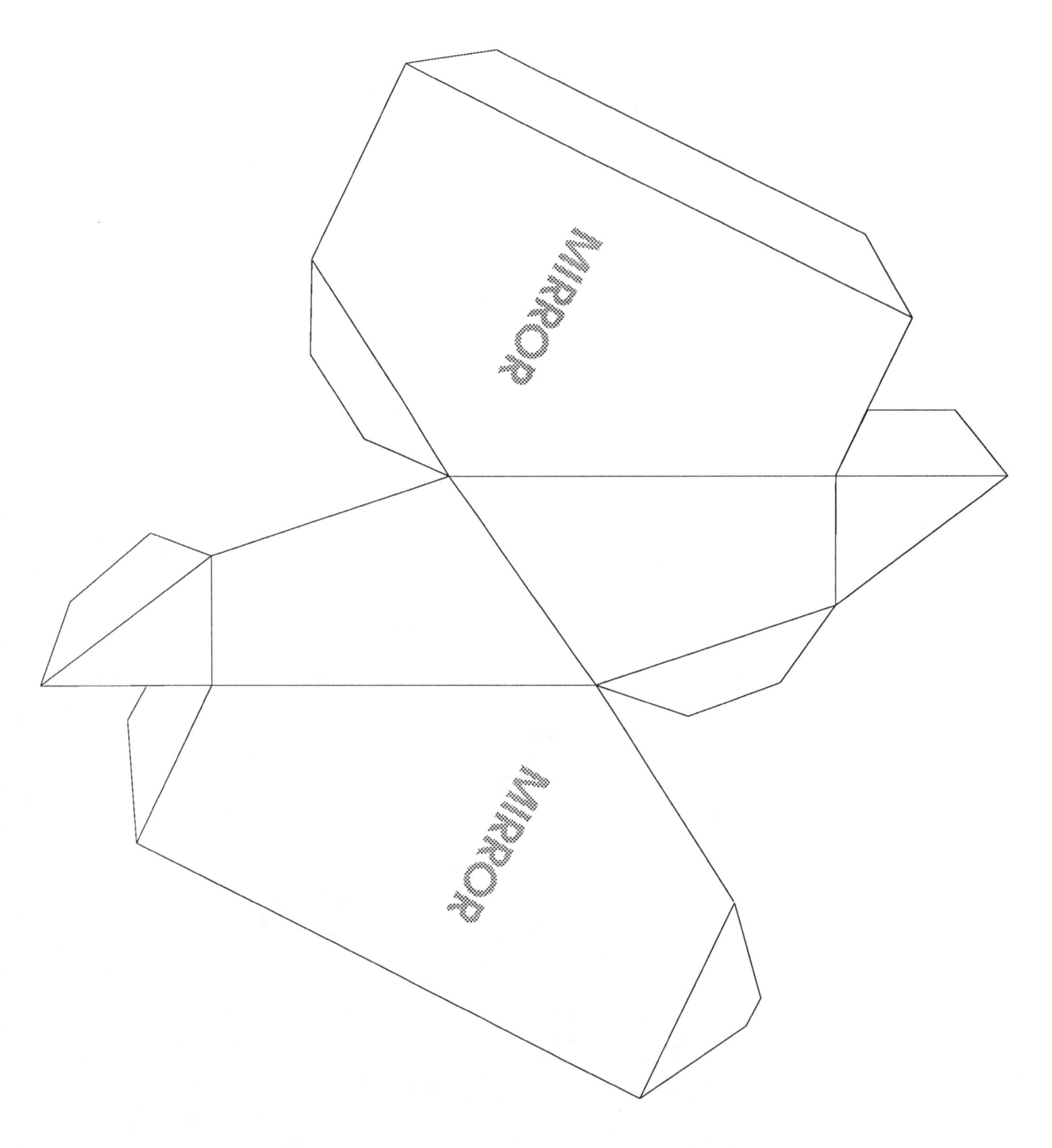

Dodecahedron Seed

Student Project 3

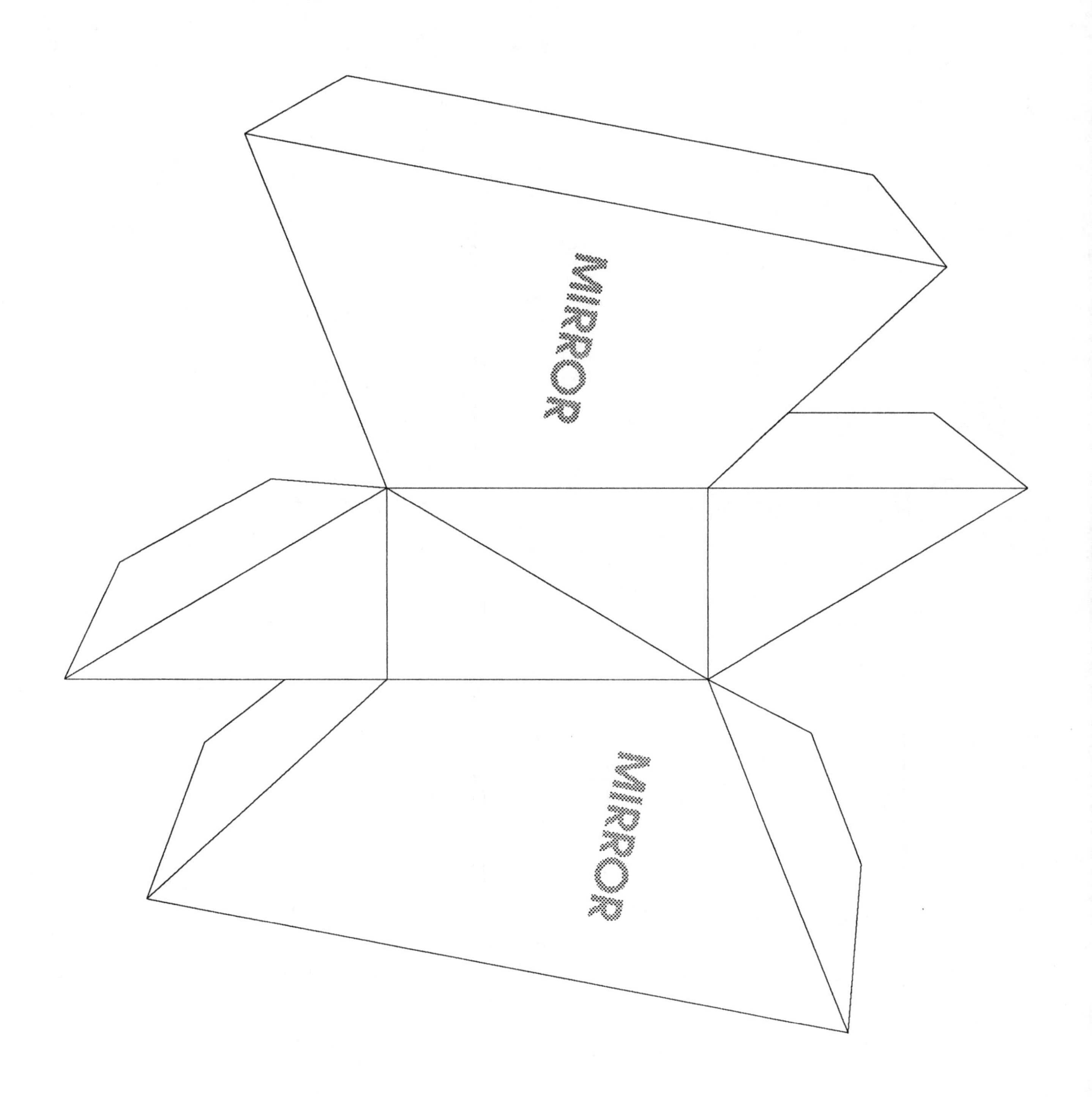

Icosahedron Seed
Student Project 3

Cube Net
Student Project 5

attach flap here

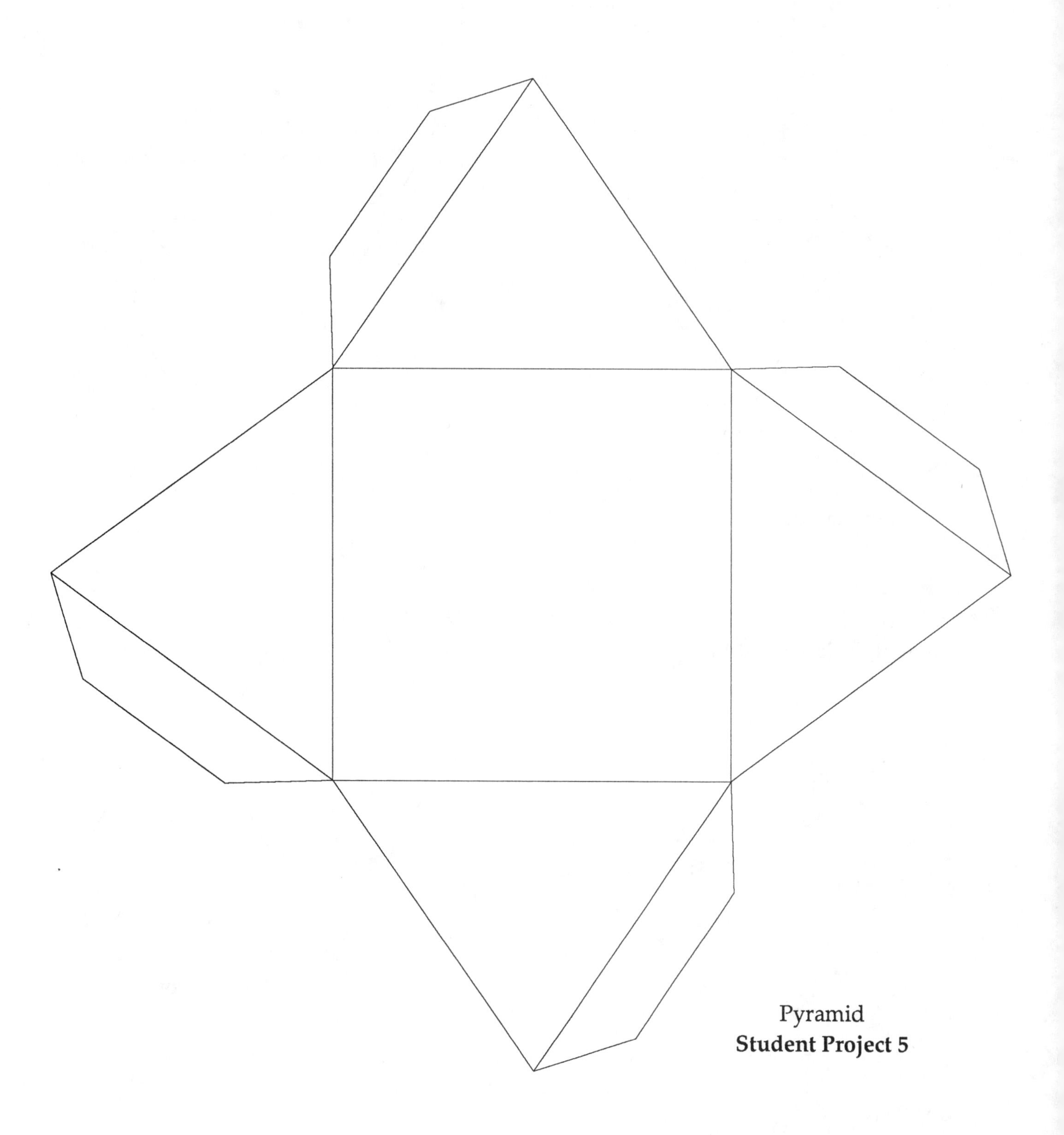

Pyramid
Student Project 5

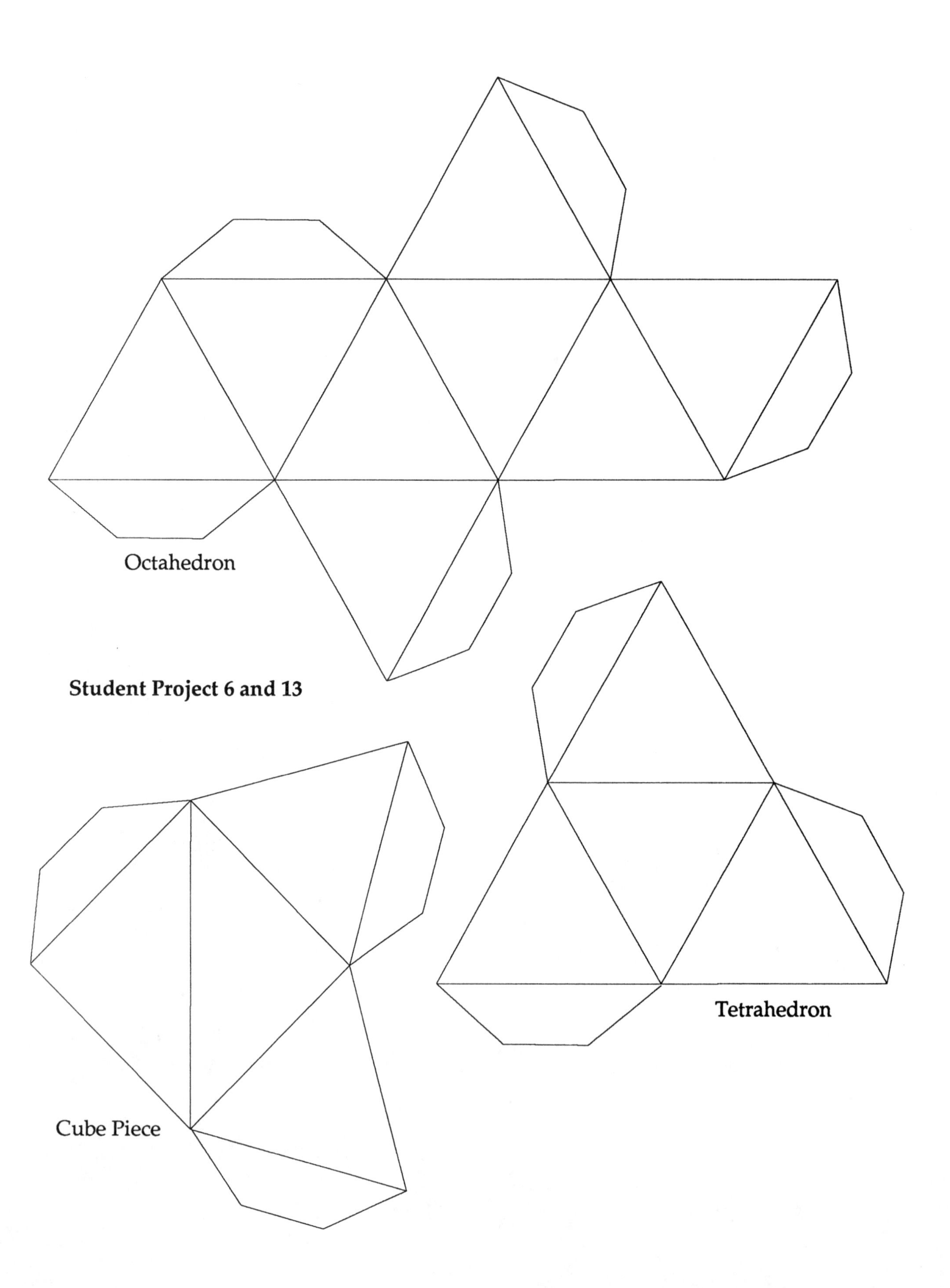
Octahedron
Student Project 6 and 13
Tetrahedron
Cube Piece

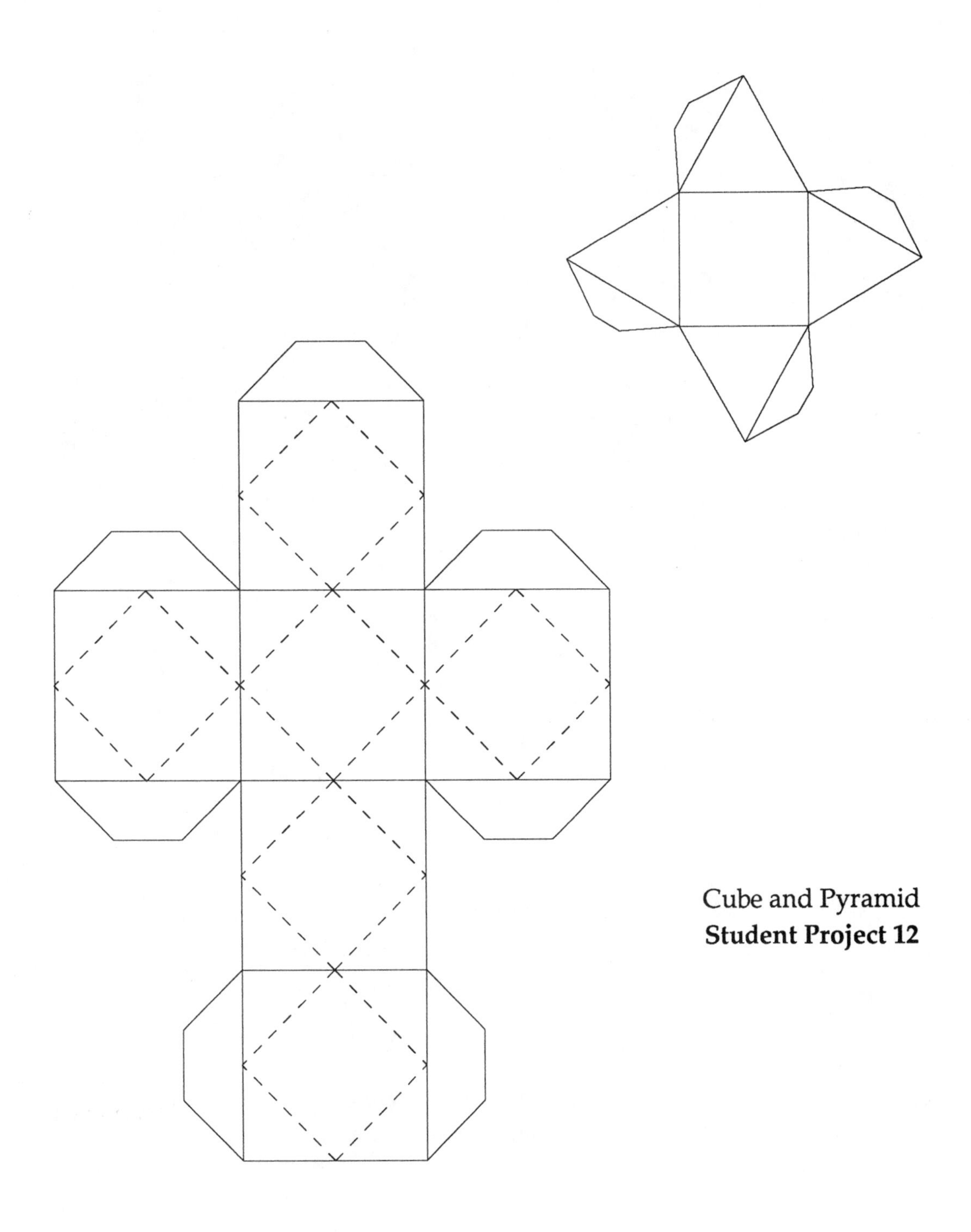

Cube and Pyramid
Student Project 12

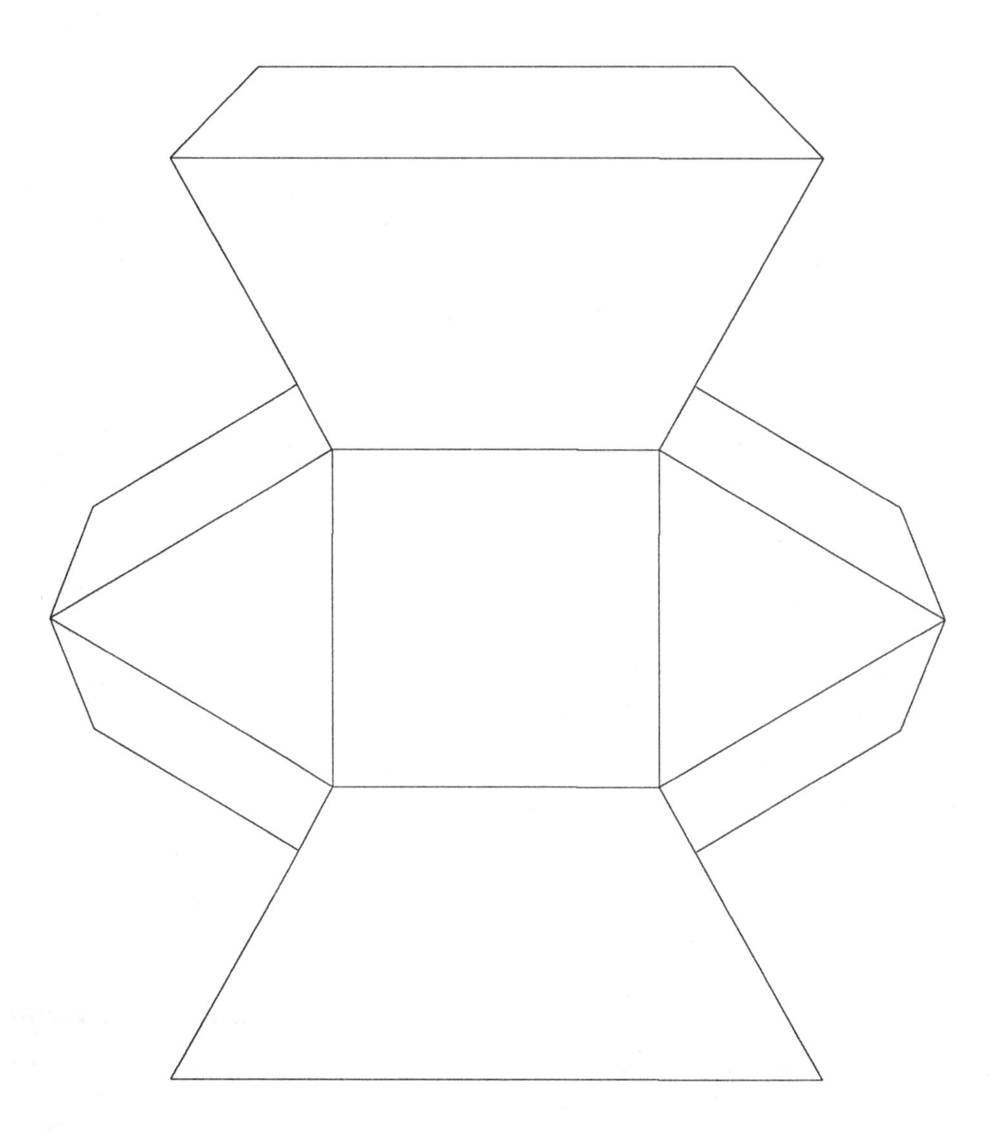

Tetrahedron Puzzle
Student Project 14